Problems for Metagrobologists

A Collection of Puzzles with Real Mathematical, Logical or Scientific Content

Problem Solving in Mathematics and Beyond

Series Editor: Dr. Alfred S. Posamentier
Chief Liaison for International Academic Affairs
Professor Emeritus of Mathematics Education
CCNY - City University of New York

Long Island University
1 University Plaza -- M101
Brooklyn, New York 11201

Published

Vol. 1 Problem-Solving Strategies in Mathematics:
From Common Approaches to Exemplary Strategies
by Alfred S. Posamentier and Stephen Krulik

Vol. 3 Problems for Metagrobologists: A Collection of Puzzles with Real Mathematical, Logical or Scientific Content
by David Singmaster

Vol. 4 Mathematics Problem-Solving Challenges for Secondary School Students and Beyond
by David Linker and Alan Sultan

Problem Solving in Mathematics and Beyond
Volume 03

Problems for Metagrobologists

A Collection of Puzzles with Real Mathematical, Logical or Scientific Content

David Singmaster

NEW JERSEY • LONDON • SINGAPORE • BEIJING • SHANGHAI • HONG KONG • TAIPEI • CHENNAI • TOKYO

Published by

World Scientific Publishing Co. Pte. Ltd.
5 Toh Tuck Link, Singapore 596224
USA office: 27 Warren Street, Suite 401-402, Hackensack, NJ 07601
UK office: 57 Shelton Street, Covent Garden, London WC2H 9HE

Library of Congress Cataloging-in-Publication Data
Names: Singmaster, David.
Title: Problems for metagrobologists : a collection of puzzles with real
mathematical, logical, or scientific content / by David Singmaster.
Description: New Jersey : World Scientific, 2015. | Series: Problem solving
in mathematics and beyond ; volume 3
Identifiers: LCCN 2015031908| ISBN 9789814663632 (hardcover : alk. paper) |
ISBN 9789814663649 (pbk : alk. paper)
Subjects: LCSH: Mathematical recreations. | Puzzles.
Classification: LCC QA95 .S496 2015 | DDC 793.74--dc23
LC record available at http://lccn.loc.gov/2015031908

British Library Cataloguing-in-Publication Data
A catalogue record for this book is available from the British Library.

Typeset by Stallion Press
Email: enquiries@stallionpress.com

Printed in Singapore

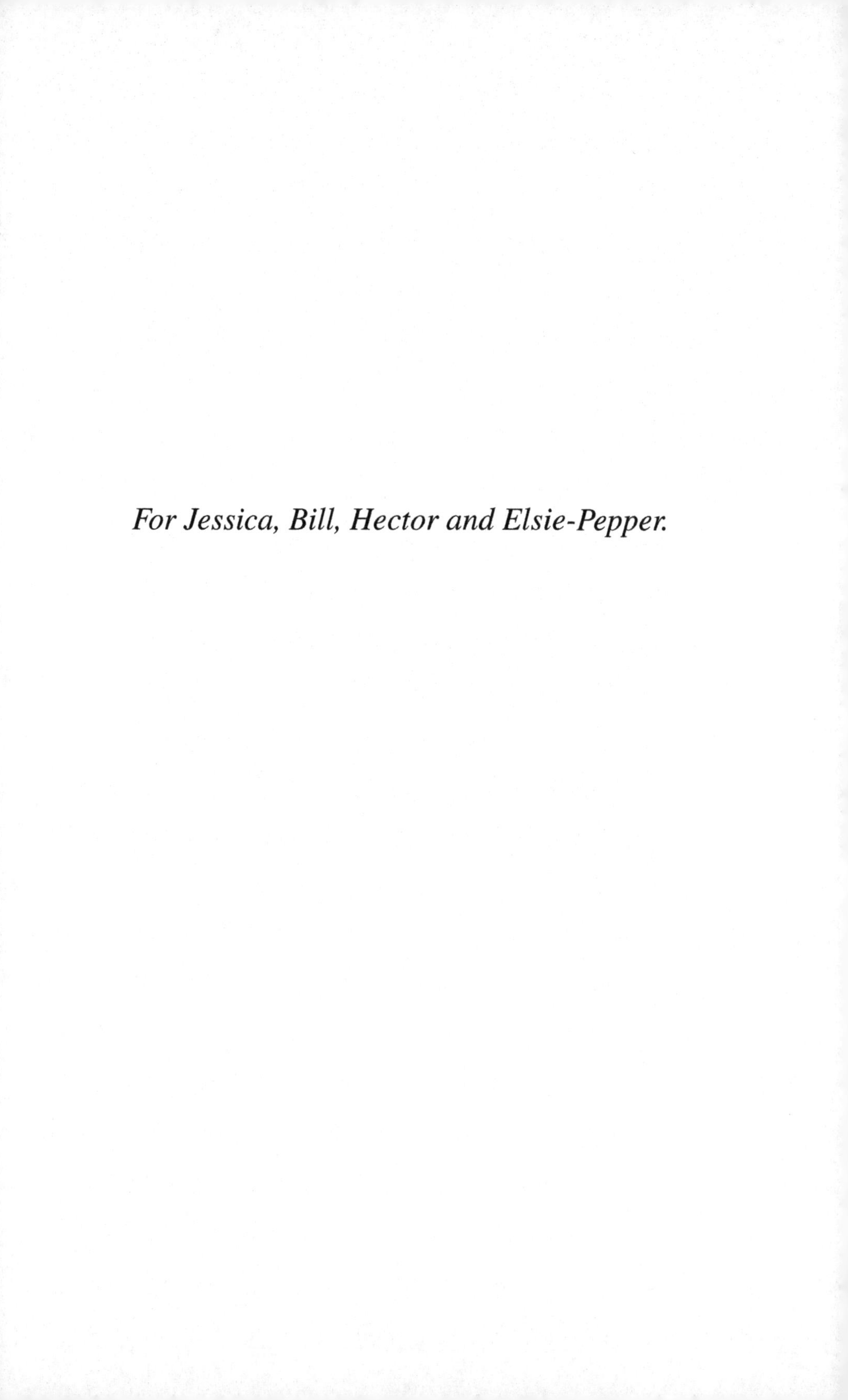

For Jessica, Bill, Hector and Elsie-Pepper.

Contents

Introduction

In case you don't already know, the *Oxford English Dictionary's* (OED) entry for METAGROBOLIZE describes it as humorous. Rabelais used *metagraboulizer* and Cotgrave (1611) translated it as "to dunce upon, to puzzle, or (too much) beate the braines about". The *OED* gives: To puzzle, mystify; To puzzle out.

Urquhart's translation of Rabelais used metagrabolizing, metagrobolism and metagrabolized. Kipling used metagrobolized in 1899, which is the latest citation given in the *OED*.

About 1981, Rick Irby, the American puzzlist, found the verb used in the *Wall Street Journal*. Since then, the noun has been adopted by a number of puzzlers as a term for one who makes and does puzzles.

Several people have suggested that the noun forms should be metagroboly and metagrobolist but metagrobology is definitely easier to say and this implies using metagrobologist.

As a mathematician and as yet unchristened metagrobologist, I contributed problems to mathematical journals since I was a graduate student in 1963. From 1987, I contributed series of puzzle problems to more popular periodicals and other media — in alphabetical order: BBC Radio 4; BBCTV; Canadian Broadcasting; *Focus* (the UK popular science magazine); *Games & Puzzles*; *Los Angeles Times*; *Micromath*; *The Weekend Telegraph* (London). This book is a collection of some of these problems, combined with material contributed to a memo pad — *Puzzle a Day* (Lagoon Books, London, 2001) — and much previously unpublished material.

My interest in puzzles has always been in understanding, not just the puzzle and its solution, but also the history of the puzzle and its generalizations. All too many puzzle books simply give the specific solution without indicating how to find it. Here I will give reasonably detailed solutions. Some solutions simply require the right insight to yield the answer almost immediately; other problems require setting up algebraic equations and carefully solving them; and some problems require checking a lot of cases and the assistance of a computer is needed. When the problem has some history, I will outline it. Some problems have immediate generalizations which can be stated and solved. In some cases, correspondents have sent me variations of the problem. In a few cases, a generalized problem is unsolved. Word-based puzzles are often rather open-ended. I will be delighted to hear from readers who find simpler solutions or better answers, create other variants of problems, or make progress on the unsolved problems.

People often ask where puzzles come from or how does one find/create puzzles. I must say 'Everywhere'. I am working on a history of recreational mathematics and I read many books containing puzzles. As the reader will see, this directly supplies a number of problems, but such problems are often open to improvement and generalization because early works give very cryptic and specific solutions. Further, a number of problems arise in everyday life if one is observant — problems about clocks, mirrors, shadows, bicycles, railway trains, letters of the alphabet, etc. Many of these have been examined in the past, but are open to generalization and variation giving new problems. Another fruitful source of problems is other metagrobologists. There is a pleasure in creating a nice puzzle which makes one want to share it. My friends rightly say that puzzle creators are essentially sadistic, like makers of puns. We want to show a neat problem to fellow connoisseurs, especially if it has a real moment of insight, an 'A-Ha' moment.

The origin of many, perhaps most, classical problems is uncertain. Typically such problems first appear fully developed but with no reference to any earlier versions. Many of these are surprisingly old, deriving from the ancient civilizations, so there is little evidence to elucidate the history. In a few cases, one can see a distinctive problem appearing in China, then India, then in the Arabic world and then in medieval Europe — but we know nothing about how the problem was transmitted along this route. In other cases, we only see the problem in two distant places, e.g. at the far ends of the Silk Road. Very rarely, we can see a problem is new at a particular time because people are having difficulty with it and getting it wrong. But even modern problems can be hard to trace — I have several problems that arose in living memory but the originators cannot be determined. Mathematical problems are like urban legends or jokes — they are part of the folklore of mathematics.

I am grateful to many collectors and creators of mathematical puzzles in the past, whose works are regularly pirated here and elsewhere, just as they pirated from their predecessors, but I will try to acknowledge them and to give new versions of their problems. Of these, the most famous are Sam Loyd, Henry Dudeney, Hubert Phillips and Martin Gardner. Gardner is by far the most important expositor of mathematical recreations of all time — he was still writing at the age of 90. He has also created some good puzzles.

Unfortunately, when I wrote some of these problems, I simply referred to, e.g. a 1930s puzzle book, and I have not [yet] relocated the source. As will be seen, puzzles occur in diverse places.

But enough waffle. Let us get down to puzzles. I will group them by type.

Chapter 1

General Arithmetic Puzzles

1. SHARE AND SHARE ALIKE

Jessica and her friend Pud like to eat a big lunch. One day Jessica brought four sandwiches and Pud brought five. Samantha got mugged on her way to school, but the mug ran off with her lunch and left her purse. So Jessica and Pud shared their sandwiches with Samantha. After eating, Samantha said: "Thanks a million. I've got to see Mr. Grind, but here's some money to pay for the sandwiches." She left $3 and ran off. Jessica said: "Let me see, I brought four and you brought five, so I get 4/9 of $3, which is 4/3 of a dollar, which is $1.33, near enough." Pud said: "Ummm, I'm not sure that's fair." Why?

Based on Fibonacci (1202)

2. LEO'S LILLIAN LIMERICK

Leo Moser was a prolific mathematician who vastly enjoyed the subject. His brother Willy kindly sent me a fine example of Leo's mathematical limericks. I (with much help from my wife) have extended this to clarify the problems.

Once a bright young lady called Lillian
Summed the NUMBERS from one to a billion
But it gave her the "fidgets"
To add up the DIGITS.
If you can help her, she'll thank you a million.

If you are as bright as this Lillian,
Sum the NUMBERS from one to a billion
And to show you're a whiz
At this kind of biz,
Sum the DIGITS in numbers one to a billion.

3. SOME SQUARES

My friends, Mr. and Mrs. Able and Mr. and Mrs. Baker, went shopping for Christmas presents. They split up and later rejoined for lunch, each having bought something. Mrs. Able said: "An odd thing happened this morning. I bought as many items as each item cost in dollars." Amazingly, each of the others said they had done the same thing, though they had all bought different numbers of items. Even more amazingly, the Ables had spent just as much as the Bakers. What is the smallest amount they could have spent?

4. SOME PRODUCT

My secretary, Ms. Flubbit, has gone to visit her relatives and Mr. Flubbit is standing in for her. He has been upholding the family tradition. I asked him to add up some positive whole numbers and he managed to multiply them instead! But when I did the sum, I got the same result as he had! I told my daughter Jessica about this remarkable occurrence. After some thought, she said: "I bet I can tell you the numbers if you tell me how many numbers there were." But I was way ahead of her and responded: "It's a good bet to try, but you'll lose!" What is the least number of numbers there could have been and how many sets of numbers could there be in this case?

5. SUM TROUBLE

Jessica and her friend Hannah were looking at a puzzle book which asked how to put + and − signs into the sequence 123456789 in order to make it add up to 100. After a bit, they looked at the answer and found $1 + 2 + 3 - 4 + 5 + 6 + 78 + 9 = 100$. Hannah said: "I bet there are more ways to do this." Jessica said: "Sure, but I don't like that minus sign, it's too complicated. I'd rather have only plus signs. I wonder if that's possible." "If it were possible, they'd have asked for it", replied Hannah. "Possibly. Or perhaps it's too easy", mused Jessica.

Who is right?

6. RUSH AND SEDGE

Here is a problem from the early Chinese classic *Chiu Chang Suan Ching (Nine Chapters on the Mathematical Art)*, which is variously dated between −2nd century (i.e. the second century BC) and +2nd century. Most of its problems are familiar and many traveled via India and the Arab world to medieval Europe. The following is an example which does not seem to have been copied elsewhere.

A rush grows 3 feet on the first day, then 3/2 feet on the second day, 3/4 foot on the third day, ... A sedge grows 1 foot on the first day, 2 feet on the second day, 4 feet on the third day, ... When are they the same size?

The *Chiu Chang Suan Ching* gives the answer 2 6/13 days. Can you see how this was obtained? Can you find a more correct solution?

7. SHE'S A SQUARE

Jessica's friend Katie says she will be x years old in the year x^2. How old will she be this year?

8. LEMONADE AND WATER

Jessica had a jug containing 7 pints of lemonade mixed with 11 pints of water. Her friend Rachel had a jug with 13 pints of lemonade mixed with 5 pints of water. Jessica poured two pints from her jug into Rachel's jug. After mixing it thoroughly, Rachel poured two pints back into Jessica's jug. They repeated this three more times. How much more lemonade has Jessica gained than Rachel has gained water?

9. AN OLD MISTAKE

A problem book from the turn of the century asks: "What is the first term of an arithmetic progression of five terms if the sum of all five terms is 40 and their product is 12320?" The book gives the answer 3. Do you agree? What question could this be the answer to?

10. HORSE TRADING

I took Jessica and her friend Hannah to a country horse fair, where we watched a trader in action. In the morning he bought a batch of similar horses. Looking them over, he decided that one was a bit better than the others and he would keep it. In the afternoon he sold the others at £20 more per horse than he had paid per horse. After some figuring, Jessica and Hannah realized that he had received just as much as he had paid out for all the horses in the morning. Jessica said: "So he would also have gotten as much money if he kept two horses and sold the others for £40 per horse more than he had paid." Hannah said: "No, I figure he would have to get £45 more per horse." I said: "One of you is right."

Who is right? And how many horses did the trader buy and how much did he pay for each horse?

11. A WEIGHTY PROBLEM

An old book of games and puzzles gives the following problem and answer.

"A man had a set of weigh-scales and only four weights, yet with those four weights he could weigh any number of pounds up to [and including] forty. What were the four weights?

The four weights were, 3 lb, 4 lb, 6 lb, and 27 lb."

Obviously our man works in a cheese factory where he wants to know how many ways he can weigh his whey. What weights of whey can he weigh? I thought that perhaps the 6 lb was a misprint for 9 lb. If he has 3, 4, 9 and 27 lb weights, is he better off and in what way?

[This problem is often known as Bachet's weight problem since Bachet described it in 1612, but it was described already by Fibonacci in 1202. Fibonacci gives a set of four weights that allows you to weigh any number of pounds up through forty pounds. You probably know the standard answer. If not, try to find such a set of weights. If you do know the answer, can you prove that it is unique?]

12. TRIANGULATION

Jessica and Rachel have been marble fanatics for the last few months. They have been playing as a team and have managed to win almost all the marbles in the neighborhood. Jessica was laying out the marbles on the floor and she started putting them into a triangular form, like the 10 pins in bowling or the 15 balls in pool or snooker. They were surprised to find that the marbles made a perfect triangle. Rachel then arranged them into smaller triangles and they were even more surprised to find that the marbles made an exact number of equal triangles. They then tried other sizes of triangles and they found eight ways altogether to form the marbles into an exact number of equal triangles. (Of course, this counted the first case when there was just one triangle.) They didn't have a huge number of marbles — certainly less than a thousand or two — so how many did they have?

13. MATCHSTICKS OR FIDDLESTICKS

By mixing Roman numerals, 1, and the arithmetic operators $+$, $-$, $\times$ and $/$, one can form various numbers with five matchsticks. For example: IV/I $= 4$, V $\times$ I $= 5$. Make all the integer values from 0 through 17. I haven't been able to make 18 — can you?

14. CALCULATED CONFUSION

Jessica likes playing with her calculator and our computer. The other day she discovered that they don't behave the same way. She put in $3 + 4 \times 5$ into her calculator and got 35, but when she put it on the computer, it gave 23. The calculator (if it is old enough) does each operation immediately, so it added 3 to 4 to get 7 and then multiplied this result by 5 to get 35. The computer looks at the whole expression and evaluates products before sums, so it multiplied 4 by 5 to get 20 and then added 3 to get 23. After I explained this to Jessica, she went away and did a lot of button punching. She came back after a bit and said she had been looking for an example where the two methods of calculation gave the same result, but she always found the first method gave a larger answer than the second method. Does this always happen? If not, when can the first result be the smaller? Can the two results ever be equal? If so, when? Why didn't Jessica find such examples?

15. JELLY BEAN DIVISION

Jessica and her neighbors Hannah and Rachel bought a big bag of jelly beans. Because they had put in different amounts of money, they wanted to count the number of jelly beans in order to share them, but they kept losing count. All they could establish was that there were less than 500. Finally Jessica decided to count them out by 7s. She found that there were just three jelly beans left over. By now they were hungry, so they agreed to eat the three. Hannah then counted out the beans by 8s. She found there were three jelly beans left over and they ate them. Rachel then counted them out by 9s and, to her surprise, she found there were three jelly beans left over and they ate them. How many beans were there at the beginning?

16. SONS AND DAUGHTERS

Reading a modern puzzle book from India, I found a somewhat cryptic version of a standard problem.

A man died leaving an estate of 1,920,000 rupees with the proviso that each son should receive three times as much as a daughter and each daughter should receive twice as much as a mother. How much did the mother receive?

Clearly some information has been lost, so I looked at the answer, which was 49,200 10/13 rupees. A little work shows that this cannot be correct. Why? What should the answer be? And how many sons and daughters might there be?

17. THREE BRICKLAYERS

I wanted a wall built in my backyard. I asked my local builder for an estimate and especially wanted to know how long it would take. He said he'd previously built several walls in the road that were just the same size. But he had three bricklayers and he'd only ever used them in pairs. When he'd put Al and Bill on the job, they did it in 12 days, but Al and Charlie took 15 days while Bill and Charlie took 20 days. So I asked how long it would take each man working alone. That stumped him but he thought he could work out how long it would take all three together but he didn't have his calculator with him. Can you find all these times for us? If you find this classic problem too easy, can you explain how to find numbers so that all the times turn out to be whole numbers?

18. AN ODD AGE PROBLEM

Jessica is just 16 and very conscious of her new age. Her neighbor Helen is just 8 and I was teasing Jessica. "Seven years ago, you were 9 times as old as Helen; six years ago, you were 5 times her age; four years ago, you were 3 times her age; and now you are only twice her age. If you are not careful, soon you'll be the same age!" Jessica seemed a bit worried and went off muttering. I saw her doing a lot of scribbling. Next day, she retorted, "Dad, that's just the limit! By the way, did you ever consider when I would be half as old as Helen?" Now it was my turn to be very worried and I began muttering — "That can't be, you're always older than Helen." "Don't be so positive," said Jessica as she went off to school. Can you help?

19. SOME SQUARE SUMS

You undoubtedly know the idea of a magic square. It's an n by n array of the first n^2 integers such that each row, each column and the two diagonals total up to the same value, which is $n(n^2+1)/2$. There is essentially just one magic square of order 3, already known to the Chinese a few centuries BC. There are 880 essentially distinct magic squares of order 4, as established by Bernard Frénicle de Bessy in about 1675.

In the construction of 4 by 4 magic squares, it is convenient to have each of the numbers 0, 1,..., 15 uniquely represented as a sum of two non-negative numbers, one number from one set and the other from another set. The easiest example is where one set is {0, 1, 2, 3} and the other set is {0, 4, 8, 12}. That is, we get the following addition table containing each of the numbers 0,..., 15 exactly once.

	0	1	2	3
0	0	1	2	3
4	4	5	6	7
8	8	9	10	11
12	12	13	14	15

Can you find any other examples?

20. A PRETTY PIZZA PROBLEM

A friend of mine managed the local pizza parlor some years ago. Every day, at the end of each shift, which was a whole number of hours, he had to pay the nine employees and the wages bill came to $333.60. There are three kinds of employees and they earned $5.00, $3.75 and $1.35 per hour. How many of each kind of employee were there and how long was a shift?

21. GEE WHIZZ!

Problem 55, part 3 in the *Whizz Kids Crazy Puzzle Book* (Macdonald, London, 1982) gives the following in an elaborate diagram on page 42.

7	3	11
9	5	22
?	2	27

What should the missing number be? On page 83, the answer is given as 19. I could make no sense of this result, and after some time, I decided that there must be a misprint somewhere. The answer is simply printed, so it could easily have been misprinted — e.g. it might have been a handwritten 10 which was misread as a 19. The problem itself is in a drawn lay-out, so it seems less likely that there is a misprint in the data, but a misplacing of a number could have been overlooked. I had this problem up on my door for some years and in October 1993, Stephen King, one of my first-year B.Sc. students, solved it. It is indeed a simple misprint and it is not difficult — one just has to hit on the right idea and I am surprised (perhaps embarrassed) not to have found it myself. Can you determine what is misprinted and what was intended?

22. SUMS OF THREE FACTORS

My colleague, William Hartston, in his delightful *Book of Numbers* inadvertently asserted that six was the only number which was the sum of three of its factors. Here we assume that these are distinct factors — e.g. $6 = 1 + 2 + 3$. Can you determine all numbers which are sums of three distinct factors? How about two, or four, factors?

23. SQUARE CONNECTIONS

Jessica and her friend Ben had been trying to find lots of Pythagorean Triples, that is, positive integer solutions of $x^2 + y^2 = z^2$. After some time, they hadn't found many solutions and, like anyone stuck with a problem, Ben suggested trying something simpler, like solutions of $x + y = z^2$. Of course there are lots of answers to this and they wrote down the first examples. Jessica observed that certain numbers repeated and could be strung together, e.g. $1, 3, 6, 10, 15, 1, \ldots$, so that the sum of each consecutive pair was a square. Ben said that one could start with 2 and go 2, 7, 9, 16, 20, 5, 4, 12, 13, 3 and this connected with what Jessica had started. Jessica looked and asked if one could make a string of all the first few numbers, but she didn't know how many numbers one would need — hundreds maybe? Can you help them out? Can you find the smallest value of N such that the numbers $1, 2, \ldots, N$ can be made into a string (not necessarily a cycle) so that adjacent terms add up to a square?

Chapter 2

Properties of Digits

24. SHIFTY MULTIPLES

Jessica's teacher, Mr. Grind, asked her to multiply 5×142857. After some time, she got 714285. But her friend Pud pointed out that all she had to do was to shift the 7 from the back to the front. Jessica thought this was a lovely short-cut, so when Mr. Grind asked her to find 9 times a number, she moved the end digit, which was a 9, around to the front and got the right answer. What was the number?

25. MORE SHIFTY MULTIPLES

You may recall that Jessica discovered that she could multiply 5×142857 by simply moving the last digit (i.e. the 7) from the back to the front and she adopted this short-cut with enthusiasm. Unfortunately she couldn't always remember which way to move the digit and when her teacher, Mr. Grind, asked her a similar problem, she moved the front digit to the back. She got the right answer. What might it be?

26. SUM MAGIC

Here is a magic trick you can play on a friend — but first you have to work it out yourself!

Ask your friend to arrange the nine digits $1, 2, \ldots, 9$ as three 3-digit numbers, ABC, DEF, GHI, so that $ABC + DEF = GHI$. This isn't very hard, so she should be able to do it fairly quickly. Now ask her to tell you what two of the digits in the sum GHI are. You immediately tell her what the remaining digit is. How do you do this?

If that's too easy, how many solutions of the above sum are there?

27. DIGITAL DIFFICULTY

Jessica was playing with her calculator and noticed that sometimes a cube had three digits, or a fourth power had four digits, etc. She tried to find all n-th powers which had n digits, but the numbers got too big for her calculator and she gave up. Can you find the largest such number? If that's too easy, find all such numbers.

28. ON THE SQUARE?

A wartime puzzle book gives the following curiosity: 648 is divisible by the square of the sum of its digits. That is, we have $6 + 4 + 8 = 18$ is the sum of the digits of 648, and $18^2 = 324$ which divides 648. The book then says there are ten such three-digit numbers. Thinking about this, I realized that the author had omitted several cases by making restrictions which he does not state. Can you find all the solutions? Can you figure out what implicit restrictions he made to get just 10 solutions?

29. SUMS OF POWERS OF DIGITS

It is remarkable that some numbers larger than 1 are equal to the sum of the cubes of their digits (in the usual decimal expansion). This idea seems to have first been raised by F. Hoppenot in a letter to the Belgian magazine *Sphinx* in 1937. Surprisingly, Hoppenot omitted one case (other than 1, which is so trivial that it is always omitted). Can you find them all?

An n-digit number which is the sum of the k-th powers of its digits is now called a PDI (Perfect Digital Invariant). We always assume that a number does not begin with a 0. If $k = n$, it is called a PPDI (PluPerfect Digital Invariant). Very little is known about PDIs — e.g. are there infinitely many of them? However, you should be able to demonstrate that there are only a finite number of PDIs for each given k and that there are only a finite number of PPDIs (though these are not all known).

30. EVEN MORE SHIFTY MULTIPLES

You may recall some time ago that Jessica discovered that she could multiply 3×285714 by simply moving the first digit to the end to get 857142 and that she could multiply 5×142857 by shifting the last digit to the front to get 714285. She is a bit older now and dealing with fractions. She wants to find numbers which multiply by 3/2 when she moves a digit from one end to the other. Can you help her out?

31. MIXED-UP MULTIPLICATION

The following occurs in a popular Russian book by Yakov Perelman, but I have never seen it elsewhere. Have you ever noticed that $46 \times 96 = 64 \times 69$? Neither had I. Can you find all such two-digit mix-ups? Having seen the solution, I naturally wondered if there are any other ways to get two-digit mix-ups? This can happen in two ways. First, Perelman has not considered zeroes — what happens if we do allow zeroes? Secondly, could the digits be permuted in some other way? The latter question is very lengthy, but straightforward. You may prefer to let a computer work at it!

32. PERMUTED PRODUCTS

Consider $14 \times 926 = 12984$. All the five digits of the terms reappear in the result. I first saw this in some series of *Ripley's Believe It or Not!*, where several examples are given and it is claimed that there are just 12 such examples. However, they have forgotten to look at other forms. There are no examples with two digits, but there are three examples with three digits. Can you find them?

33. RUBBED OUT!

$$\begin{array}{r} 111 \\ 333 \\ 555 \\ 777 \\ +999 \\ \hline 2775 \end{array}$$

Consider the addition sum above as though it were on a blackboard or a slate. An old problem, probably from the days when slates were used, is to rub out some of the digits to leave a sum which totals 1111. The problem is usually given with a number of digits to rub out, but a version I recently saw showed examples with 5 and 9 digits rubbed out — e.g. $111 + 333 + 500 + 077 + 090$ — and then asked to have 6 digits rubbed out, saying it wasn't so simple. After working this out, I wondered: "Which numbers of digits can be rubbed out to leave a sum of 1111?" and then I asked how many solutions there were. It takes a little work, but I am sure you can solve these problems.

34. NEW CENTURY COMING UP

A popular catch puzzle of the 19th century was "Arrange the figures 1 to 9 in such order that, by adding them together, they amount to 100." The answer was: $15 + 36 + 47 = 98 + 2 = 100$. There are a number of similar answers, but one cannot get a pure addition sum of all the digits adding to 100 because of a generalized parity argument ('casting out nines' or (mod 9) arithmetic) — as discussed in Solution 5. One can get quite simple answers if one does permit subtraction — e.g. $1 + 2 + 3 - 4 + 5 + 6 + 78 + 9$ or $123 - 45 - 67 + 89$. Suppose one just uses the digits $1, 2, \ldots, n$, with $n < 9$. For which n can one get a pure addition to 100? Can one use the digits $1, 2, \ldots, n$ to make 100 in some other way?

35. THE RATIO OF A NUMBER TO THE SUM OF ITS DIGITS

The digital sum, or sum of the digits, of a number occurs in several ways in studying numbers. An 1897 algebra text gives a problem where a number is divisible by its digital sum, with other conditions. This led me to study those numbers which are divisible by their digital sum. Clearly this holds for any one digit number. Can you find the two-digit numbers with this property? This can be done by brute force direct searching, but can you find a logical approach?

The same algebra book gives a problem where an integer N is divisible by the sum of its digits, S, and by the product of its digits, P. Can you find all the two-digit examples?

This led me to ask if $N = SP$ ever holds. It clearly holds for $N = 0$ and $N = 1$, but I have found two other examples, which are not very large. I made a computer search to 10,000,000 and found no other examples. It is not hard to show that such numbers must have at most 60 digits and my colleague Tony Forbes has skilfully extended the computer search and determined there are no other examples.

36. A DIGITAL CURIOSITY

A book gives an example of a four-digit number *ABCD* such that if we split it into two 2-digit numbers *AB, CD,* then the square of $AB + CD$ is *ABCD.* The book then asks for another example, implying that there is just one other example. However, a little computer tells me there are more solutions than that.

37. TWO AND TWO ARE FOUR, OR FIVE

An old rhyme goes: "Two and two are four, Four and four are eight, Eight and eight are sixteen, Sixteen and sixteen are thirty two, Two and two are four, . . . ". At the last step, we are just considering the units digit of 32, or, in modern terms, we are working (mod 10). If we also do this at the previous stage, we have a cycle of doublings: 2, 4, 8, 6, 2, 4, ... Now this process cannot cycle through all the digits as we can only have even numbers as a result of a doubling. So suppose we allow a choice of the double or the double plus one at each stage. For example, from 2, we can go to 4 or to 5; from 8, we can go to 6 or to 7. Can we then find a cycle of all ten digits? How many such cycles are there?

38. PRIME PAIRING

In the sequence 11317197, each pair of consecutive digits is a two-digit prime number and these primes are all distinct. Find the longest such sequence. How many such sequences are there?

Chapter 3

Magic Figures

A magic square is a square arrangement of the numbers $1, 2, \ldots, n^2$, such that the sum of the numbers in each row, column and diagonal is the same. Naturally, there are many ways that these numbers $1, 2, 3, \ldots, n^2$ can be arranged to meet this requirement of equal sums. The most famous and oldest example is the 3×3 magic square, apparently from ancient China, around the year 0, shown below, where $n = 3$. Older Chinese writings claim this dates back to the legendary Emperor Yu, about 2200 BC, and attribute various mystical meanings to it. For this magic square, each row, column and diagonal has a sum of 15. There are many ways to achieve such a magic square, one of which is offered here (Figure A).

6	1	8
7	5	3
2	9	4

Figure A

For the 3×3 magic square, we can rotate and/or reflect the solution into 7 other solutions, but we consider these 8 solutions to all be equivalent. There have been many problems posed over the centuries regarding the placement of numbers in various patterns that would yield desired sums. In this section we will consider examples of several types of number arrangements taken from older works, where often the solutions provided there were incomplete — it often requires a computer to be sure that all possible solutions have been found.

39. SUM SQUARED

Jessica and her friend Henrietta were working on their sums. Jessica was doing the following:

$$\begin{array}{r} 16 \\ +32 \\ \hline 48 \end{array}$$

Henrietta looked over Jessica's shoulder and remarked, "All the digits are different. That's nice." Jessica looked at it for a minute and said, "You know, if you add them across, the results are still all different." "Sorry, I don't see what you mean." Jessica then wrote out the following.

$$\begin{array}{ccccc} 1 & + & 6 & = & 7 \\ + & & + & & + \\ 3 & + & 2 & = & 5 \\ = & & = & & = \\ 4 & + & 8 & = & 12 \end{array}$$

"See, when I add across each row, I get the results 7, 5 and 12 which are all different from the six numbers I had earlier."

"Wow, that's really nice," said Henrietta, "All nine numbers are different." After a minute she continued, "It would be really *really* nice if the nine numbers were the nine digits: 1, 2, . . . , 9. I betcha there must be some way to do that, because it would look so pretty."

Should Jessica take this bet or not? Suppose we ignore the lower right corner (i.e. 12 in the above example), how many examples are there?

40. USING ALL THE DIGITS

An old puzzle book asked for arrangements of the nine positive digits into three 3-digit numbers having the proportion 1 : 2 : 3. I did this by careful trial and error. I then wrote a little program to check my work and it gave an extra answer. This was because I forgot to check that certain digits were not 0. The extra answer is an arrangement of nine of the ten digits into three 3-digit numbers having the proportion 1 : 2 : 3. Amending my program, I found some more of these. Can you find all such arrangements?

41. A MAGICAL CROSS

```
        2
        4
    1 3 5 7 9
        6
        8
```

Jessica and her friend Sarah were fiddling with part of a deck of cards. They had the ace through 9 of one suit and they put out the odd and even cards as shown above. Jessica thought about it for a minute and said that the row and the column added up to the same value, namely 25 — of course she counted the ace as a 1, as in the figure. Sarah thought about this and said: "That's magic! Can we do it any other ways?" Jessica said: "There are lots of ways. See, I can just exchange the 1 and the 3. But that's pretty obvious. Or I can exchange the odds and the evens, leaving 5 in the middle, but that's obvious also. But I can exchange the 1, 9 pair with the 2, 8 pair and that really gives a different solution." Sarah said: "How many solutions are there? There are lots of pairs that add to 10 and we can re-arrange them in several ways. Will that give all the different solutions?" Jessica replied: "What's so special about 25? Can we have any other magic sums? How many solutions do they have?"

42. GOING ROUND IN CIRCLES

A recent puzzle book from India asks if you can arrange the 10 numbers $1, 2, \ldots, 10$ in a regular circle so that the sum of any two adjacent numbers is the same as the sum of the numbers opposite to them. Only one answer is given, but there are more. Can you find all of them? Can you do this with $1, 2, \ldots, 8$?

43. A MAGIC HOURGLASS

You are familiar with the 3 by 3 magic square, described at the beginning of this chapter. I recently found the following 'hourglass pattern' in a book. One wants to fill the seven positions with the numbers $1, 2, \ldots, 7$, so that both horizontal lines and the three lines through the center all add to the same value. Where I saw it, this value was given and only one solution was given, but I think you are clever enough to find out what values are possible and to find all the solutions.

$$\begin{matrix} A & B & C \\ & D & \\ E & F & G \end{matrix}$$

44. THE MAGIC OF 67

My neighbor at number 67 was a craftsman with mathematical interests. He carved two stone plaques for his gateposts. One has 67, sixty seven and LXVII carved in various styles. The other has the following square pattern which you will all instantly recognize as a magic square with constant 67.

16 19 23 9
22 10 15 20
11 25 17 14
18 13 12 24

I'm sure you have also noticed that this is not a 'consecutive' magic square, because it does not use consecutive numbers. Can you show that it is impossible to have a 4 by 4 consecutive magic square with constant 67? However, he has done the next best thing in that he uses 16 numbers from a range of 17 — that is, the sequence of numbers used only skips one number. We can call this an almost-consecutive magic square. For which house numbers can we find an almost-consecutive or a consecutive 4 by 4 magic square to give the number?

[Sadly, my neighbor has died. The new owners of the house have rebuilt the front and placed the plaques at the bottom of the front wall.]

45. MAGIC TRIANGLES

My daughter Jessica and her friend Ben had been studying magic squares at school. At home they decided that squares were square, i.e. boring, and so they tried to make magic triangles. After some experimentation, they decided to look at a pattern of six numbers, like that below, such that the sum along each edge was the same. Of course, to be magic, they used the first six positive integers, once each, as in a magic square. They found some solutions and then they couldn't find any more. They asked me if there were any more and I soon found that they had indeed found all the solutions. Can you do as well?

$$\begin{matrix} & A & \\ B & & C \\ D & E & F \end{matrix}$$

46. A DIFFERENT MAGIC TRIANGLE

After Jessica and Ben had found all their magic triangles (like the pattern below), they showed them to their friend Gwain. Needless to say, he wanted to try something even more different, so he said: "What's so magic about adding things up? Let's be different and look at the differences." "Huh?" said Jessica and Ben in unison. "Look, if you have three numbers along an edge, like 1, 2, 3, then the number in the middle is the difference of the numbers at the ends. So you want to put the numbers 1 through 6 on the triangle so this works along each edge." They set to work and soon found a solution but got stuck and couldn't find any more, but Gwain thought they hadn't tried every case. Can you find a solution? Are there any more solutions?

$$\begin{matrix} & A & \\ B & & C \\ D & E & F \end{matrix}$$

47. MAGIC CIRCLES

Draw three interlocking circles. These intersect at six crossing points *A*, *B*, *C*, *D*, *E*, *F*, as shown below.

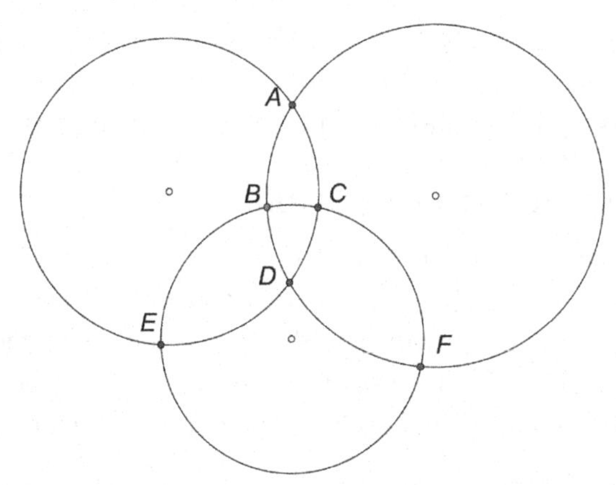

Each circle passes through four of these intersection points, and A, C, D, E; A, B, D, F; E, B, C, F are these quadruples. I recently saw a puzzle book which placed the digits 1 to 5 on five of these points and asked how to place the sixth so the sum of the values on each of the circles was the same. This is much too easy for you readers, so instead I will ask you to describe ALL the ways one can place the first six digits on these six points so that the sum of the values on each of the circles is the same.

48. ANOTHER MAGIC CIRCLE

A 1965 puzzle book from America asks you to arrange the first 11 numbers regularly on a circle with one number in the middle so that the sum of the three numbers on each diameter is a constant. That is, we want a simple kind of magic circle. Only one example is given. Can you describe all examples?

49. A MAGIC FRAME

A 1927 puzzle book gives a kind of magic figure using playing cards. After a little examination one sees that it is asking for a magic frame in the shape of a rectangle. Writing it as shown below, we want $A + B + C + D = D + E + F = F + G + H + I = I + J + A$, where the values are the integers $1, 2, \ldots, 10$. The author of the 1927 puzzle book gives one solution and sort of says it is unique. Is he correct??

$$\begin{matrix} A & B & C & D \\ J & & & E \\ I & H & G & F \end{matrix}$$

[This also appears in Bill Severn; *Packs of Fun: 101 Unusual Things to Do with Playing Cards and To Know about Them*; David McKay, NY, 1967, p. 84, asking for solutions with total 18 and giving just one answer.]

50. A MAGIC TRIANGLE SQUARED

Many mathematicians of note have composed mathematical puzzles, but only a few have produced books of puzzles. One of the most unexpected is Giuseppe Peano, the great worker in the foundations of mathematics. He wrote *Giochi di Aritmetica e Problemi Interessanti* (G. B. Paravia, Torino) in 1924. The following problem is based on a result in that book.

Imagine a triangle with four positions along each side, as shown below. To be magic, we want to put the integers 1, 2, ... , 9 in these positions so that the sum of the values along each edge is a constant. It's not too hard, but a little tedious, to find all the solutions. To prevent finding different forms of the same triangle, let us agree that $A < D < G$, $B < C$, $E < F$, $H < I$. Without this, we would get each basic solution 48 times!

Both Peano and another puzzle book from the 1920s give an example of the above where the sums of the squares of the values along a line is constant, but they are not clear as to whether there might be more examples. Can you find one such example? Are there more?

$$
\begin{array}{ccccccc}
 & & & A & & & \\
 & & I & & B & & \\
 & H & & & & C & \\
G & & F & & E & & D
\end{array}
$$

51. YET ANOTHER DIFFERENCE TRIANGLE

A few problems ago, I asked you to find triangles using the first six digits where the digit in the middle of each row was the difference of those at the ends. Andrew Healy, of Ashford, Middlesex, misread this as asking for each of the upper entries to be the difference of the digits below it. That is, he wanted:

$$A = B - C \quad \text{or} \quad C - B; \quad B = D - E \quad \text{or} \quad E - D;$$
$$C = E - F \quad \text{or} \quad F - E.$$

He found one solution and wondered if it was unique. Can you help him out?

$$
\begin{array}{ccccc}
 & & A & & \\
 & B & & C & \\
D & & E & & F
\end{array}
$$

52. STILL ANOTHER DIFFERENCE TRIANGLE

Another reader, G. C. Mudd, of Huddersfield, misread the difference triangle problem in a different way. He tried to find a kind of magic difference triangle of the first six digits such that $A+D-B = A+F-C = D+F-E$. In fact, he found all the answers, except he overlooked one case. Can you find all answers?

$$\begin{array}{ccccc} & & A & & \\ & B & & C & \\ D & & E & & F \end{array}$$

53. A DICEY MAGIC SQUARE

The Boy's Book of Carpentry and Electricity, an undated book of the 1930s(?), suggests making nine identical cubical dice and arranging them in a 3×3 array so that each row, each column and both diagonals add up to nine. It then says "The solution is as follows:

4 2 3
2 3 4
3 4 2"

By now, you should be treating such statements with some skepticism. Indeed you should immediately see two reasons why this statement is wrong. How many distinct solutions are there, assuming that solutions that differ by rotation and/or reflection are the same? [You may want to consider solutions containing a 6 and then those containing a 5.]

54. ANOTHER MAGIC FRAME

In a recent copy of *SYMmetry Plus* (a UK magazine for young mathematicians) there is a problem of using four dominoes from a double-six set to make a hollow 3×3 frame such that each of the edge triples has the same sum. It is easy to find examples, but the example given in Figure B has some of the values being the same. If one uses a larger domino set (e.g. a double-nine set), one can produce examples where all the numbers are different. This could allow us to consider this shape as a magic figure. That is, can we put the numbers 1, 2, . . . , 8 into the cells of the figure (Figure C) so that the four edge triples have the same sum. Despite the simplicity of this question, I don't recall seeing it before.

5	1	3
4		3
0	6	3

Figure B

A	*B*	*C*
H		*D*
G	*F*	*E*

Figure C

55. A GENERALIZED MAGIC SQUARE

A	B	C
D	E	F
G	H	I

I recently found this problem in a 1930s German puzzle book. We want to put the nine positive digits into a square array, as above, like a magic square, but with different sums. We want the edges and the diagonals to add up to 18, i.e.

$$A+B+C=G+H+I=A+D+G=C+F+I$$
$$=A+E+I=C+E+G=18.$$

A little work shows that then the midlines must add up to 9, i.e.

$$D+E+F=B+E+H=9.$$

The author gives just one solution. Are there others?

It struck me that the question could be generalized by letting the edges and diagonals add up to S and the midlines add up to T. One sees that $2S+T=1+\cdots+9=45$, so that T must be an odd number. Careful consideration of cases or a small computer program will find all such solutions and I was surprised by the result.

REFERENCE

Mitis, Caesar. *Rechnerische Scherze Zahlenkunststücke und Geometrisches für Jung und Alt.* Verlag von Otto Maier, Ravensburg, nd [1930s?].

56. THE MAGIC OF SEVEN

```
A – B – C
|  \ |  /
|    D
|  / |  \
E – F – G
```

In the diagram above, we want to insert the integers $1, 2, \ldots, 7$, so that all the indicated lines have the same sum S. That is, we want

$$A + B + C = A + D + G = E + D + C = E + F + G = B + D + F.$$

How many ways can this be done?

An extension of: Alan Ward; *Simple Science Puzzles*; [From Science Activities, US, 1970–1973]; Batsford, 1975, pp. 23 & 25.

Chapter 4

Monetary Problems

Since I have lived in England for many years and these problems were written for English readers, I have often used old English currency: pounds, shillings and (old) pence. A pound (£, from the Latin *libra* for a pound weight) was 20 shillings and a shilling (s from the Latin *solidus*, a Roman gold coin) was 12 pence (d, from the Latin *denarius*, a Roman silver coin). After decimalization in 1971, the British coins were $\frac{1}{2}$, 1, 2, 5, 10, 20, 50 pence (p). With inflation, the $\frac{1}{2}$p coin disappeared and the £1 and £2 were introduced.

In British English, 'check' is spelled 'cheque', from the older usage of Exchequer for Treasury, so called because accounts used to be done on a chequerboard.

57. FUNNY MONEY

When I first came to England, pounds, shillings and (old) pence were still in use. Being a stranger, I wasn't quite sure how to write down an amount and I sometimes got the pounds, shillings and pence confused. I did this one day and later noticed that my confused notation represented the correct amount of money. What is the least amount this could have been? (Of course, the amounts of pounds, shillings and pence were all different, otherwise things aren't confusing enough.)

58. NO CHANGE

Jessica asked her friend Sarah for change of a pound. Sarah opened her purse and said: "I've got lots of coins here. There aren't any pounds, but there's well over a pound." After some counting, Jessica said: "But you can't actually change a pound." What is the most money Sarah could have had? [For those outside the UK, the current UK coins are 1p, 2p, 5p, 10p,

20p, 50p. If you want to try the US version, the current coins there are 1, 5, 10, 25 and 50 cents. Canada is the same as the US except they have no 50¢ pieces.]

59. ALL CHANGE!

Jessica asked her friend Sarah for change for a pound. When she looked at the result, she was intrigued to see that she had some of each possible coin. [For non-British readers, the current coins are 1, 2, 5, 10, 20 and 50 pence.] When she said this to Sarah, Sarah said that she could have given change in several other ways, still using each current coin. Jessica then asked how many ways this could be done. Can you help out? What money would Sarah have to have in order to be able to give change in any one of these ways? [For the more ambitious, how many ways can one make change for a pound in general?]

[For the US, the current coins are 1, 5, 10, 25, 50 cents. Canada has the same except it has no 50¢ pieces.]

60. NO CHANGE!

I recently wanted to pay someone for an item that cost less than £5, so I offered him a £5 note. He said he couldn't give me the right change. So I jestingly offered him two £5 notes and was surprised when he said that would do nicely. How can this happen?

Can this happen in America?

61. AN OLD MONEY PROBLEM

What amount of pounds and shillings has the property that if you interchange the amounts of pounds and shillings, you get twice as much? For those who don't remember, or never knew, a pound had twenty shillings in the old days and a normal sum had less than 20 shillings in it. Now why have I insisted on using old money for this problem?

62. A PROFITABLE ERROR

Jessica went to the bank to cash a cheque. The pressure of business was such that the teller inadvertently interchanged the pounds and the pence in giving her money and Jessica didn't notice until sometime later. She had come out ahead, but she wondered how much she could have come out ahead. In particular, what are the maximum and the minimum profits she could make?

63. A PROFITABLE RATIO OF ERROR

Recall that when Jessica went to the bank to cash a cheque, the teller inadvertently interchanged the pounds and the pence in giving her money and Jessica didn't notice until sometime later. She came out ahead and we previously found the maximum and the minimum profits she could make. But what are the maximum and minimum of the ratio between the two amounts?

Chapter 5

Diophantine Recreations

Diophantus lived about 250 AD at Alexandria. Almost nothing is known about him, but he wrote a book, *Arithmetica*, which was a milestone in the development of algebra. In this, he solved many problems, but he was perhaps the first person to ask for all possible solutions — previously people had generally been content with one solution. His work was first printed in Greek and Latin in 1620. (It was in Fermat's copy of this work that Fermat wrote the famous marginal note now called his "Last Theorem"; Fermat's son published an edition with his father's annotations in 1670, but the original copy was later lost in a fire.)

Since then, problems where there are many, even infinitely many, integral answers have been generally called Diophantine Problems. Here are some examples.

64. BYZANTINE SALESMANSHIP

A Venetian merchant was dying and he called his nine sons to his bedside. He gives the eldest 90 pearls, the next eldest 80 pearls, etc., down to the youngest who gets only 10 pearls. The pearls are finely matched, so they are all of equal value. He instructs them to go and sell them in the markets at Padua and Venice and to divide his fortune proportionally to the amount they earn. The sons go to Padua, and being fair-minded, they all resolve to sell their pearls at the same price. They all sell some pearls and have some left over which they take to Venice. Again they all resolve to sell their pearls at the same price. Despite these complications, they all earn the same amount. How did this happen?

(Based on a problem of Niccolò Tartaglia (1556).)

65. TESTING TIMES

I recently saw an examination timetable where "6 $2\frac{1}{2}$-hour exams" was mistyped as "62 $\frac{1}{2}$-hour exams". The first occupies 15 hours in total while the second occupies 31 hours. Can the two total times ever be the same?

66. GREEK PUZZLE BOXES

According to Plutarch, the Greeks found those rectangles, with integer sides, whose area is equal to their perimeter. Can you find them?

What rectangular boxes, with integer sides, have their volume equal to their surface area?

67. A MIDDLE EASTERN MUDDLE

An Arab sheik had three sons. He died and his will left his oil wells to be divided among them. The eldest son was to receive 1/2 of them, the second son 1/3 and the third son 1/7. When they went to divide the wells, they discovered there were 41 wells. Since there is no satisfactory way to divide a well, they were puzzled as to how to proceed. They called on their Uncle Omar who was wise but poor — he only owned one oil well! How did he solve the problem?

If you know the problem already, then determine how many triples of reciprocals of integers $1/a$, $1/b$, $1/c$ can be used in such a problem — assuming $a \leq b \leq c$?

68. MORE PUZZLE BOXES

You may recall that I asked you to find those rectangular boxes with integer sides whose volume and surface area were equal. As usual, my daughter Jessica misread this and she used the sum of the edges instead of the surface area. She didn't get very far with this, so she decided to look for boxes where the sum of the edges is equal to the surface area. She got just as far with this, which was more successful, but in neither case did she think she had found all the solutions. Can you help her out? Why was she more successful in the second case?

69. PISTOLES AND GUINEAS

In the 18th century, British coinage included guineas, which were gold coins supposed to be worth one pound, but they were made of gold from Guinea in Africa and were somewhat purer than other gold coins and were consequently valued at 21 shillings. There were also other coins in circulation, notably the Spanish pistole, valued at 17 shillings. A 1745 text asks how I can pay £100 if I only have guineas and pistoles.

70. ALMOST PYTHAGOREAN

Modern mathematics uses many objects in the classroom to encourage the creativity of the student. In fact, students can often come up with problems that are too hard for their teachers to answer! The following is one such problem which arose from using Cuisenaire rods which are sticks of integral lengths.

Student: "Miss, I've made a right triangle. At least it looks like a right triangle. It has sides 5, 5 and 7."

Teacher: "It does look like a right triangle, but how would you check that it really is a right triangle?"

Student: "I guess I'd have to see if it satisfies the Theorem of Pythagoras."

Teacher: "Quite right, a triangle is a right triangle if and only if it satisfies the Theorem of Pythagoras. Does yours?"

Student: "Not really — I've got $5^2 + 5^2 = 25 + 25 = 50$ while $7^2 = 49$."

Teacher: "That's as close as you can get with integer lengths — you have: $x^2 + y^2 = z^2 + 1$."

Student: "We could also have $x^2 + y^2 = z^2 - 1$. I think we should call these 'almost Pythagorean triangles'. I think they're really neat. How can we find some more? Let's see, we don't really want any sides of length 0, though 0, 0, 1 works in the equations. Aha! We can always take $x = z$ and $y = 1$, but that gives us an isosceles triangle and it's hard to imagine that it's almost a right triangle. So I think we want to have all sides distinct and greater than 1. Ummm."

At this point, the student needs some help. Indeed, so does the teacher! In fact they found a few examples but were not sure if they had found the simplest one. Can you find any examples? Can you find the simplest example? Can you find all examples?

71. ALMOST PYTHAGOREAN TRIANGLES ON A GEOBOARD®

In the last problem, a student raised the idea of almost Pythagorean triangles, i.e. triangles with $x^2 + y^2 = z^2 \pm 1$. Some days later the student was playing with a Geoboard®. This is an array of pins arranged in a square lattice, like the intersections on a chessboard. Students can put rubber bands over the pins to make various polygonal shapes. Naturally the student wanted to make an almost Pythagorean triangle on her Geoboard®, but she couldn't seem to find one that would fit. She wondered if this was because her Geoboard® wasn't big enough or she just hadn't looked hard enough. Can you help her out?

72. GEOBOARD® TRIANGLES AGAIN

Recall that a Geoboard® is a square array of pins on which one can place rubber bands to form polygons. A school teacher asked me how many different triangles could one make on an $n \times n$ Geoboard®. After working on this for a while, I realized that there might be congruent triangles which I wasn't recognizing. The Geoboard® has natural motions which move it to itself: translations horizontally and vertically; rotations by 90°, 180° and 270°; reflections in the horizontal bisector, the vertical bisector and in the two diagonals. If a triangle can be moved onto another triangle by some motions of the board, we will say that they are (Geoboard®) equivalent. After some fiddling, I found an example of a triangle that could be formed in two ways on the Geoboard®, but these ways were not equivalent. A triangle with sides 15, 20, 25 can be placed with its vertices at (0, 0), (15, 0), (0, 20) and also with its vertices at (0, 0), (25, 0), (16, 12) and this is the simplest example where we have a right triangle with the legs along the axes in one position and with the hypotenuse along an axis in the other position. Can you find some simpler examples which do not have the special form of my example?

73. THE CCCC RANCH

Traveling through California, one expects to run across odd things, but I was more than a bit intrigued to see a sign post: THE CCCC RANCH. So I drove off along the indicated side road and about five miles later I found the ranch, which was really more of a small farm. A young man came out and asked our business. When I told him it was just curiosity about the name, he guffawed, "Lots of folks wonder about that. Well, we raise four kinds of animals that begin with C." "I see," I replied, "if you'll excuse the pun, a cow and a chicken already. What are the other kinds?" "Children and copperheads — we found so many of the damn snakes on the land that we've gone into raising them for their venom."

My daughter Jessica piped up, "How many snakes have you got?" "Do you like puzzles?" "Oh, yes!" "Well then, at the moment we've got 17 heads all together." "Do you have any two-headed animals?" "Nope, but that's a smart question. The animals have 11 tails." "Do you count people as having tails?" "Nope, these are all genuine tails and there ain't any double tailed beasts either!" "How many legs have they got and are they all normal?" "They've got 50 legs and they're all normal." "Let me see," said Jessica, "there are four unknown numbers, so I need some more information." "Sorry," said the lad, "that's all I'm gonna tell you, except that we've got some of each."

Can Jessica find the number of animals of each kind?

74. DOUBLING UP

Jessica and Sophie were playing poker with matchsticks. Each time, they bet as much as possible, namely as much as the poorer one had. Jessica won first, then Sophie, then alternately through six hands all together. At this point, they were amazed to notice that they both had the same number of matchsticks. What are the least numbers of matchsticks they could have had at the beginning?

[If that's too easy, solve the problem for n hands.]

75. A STRANGE CHESSBOARD

Maestro Biagio, an early 14th-century Florentine teacher, notes that an ordinary 8 by 8 chessboard has 28 squares along the edges and 36 squares in its interior. He then asks what size of chessboard has the same number of edge squares as interior squares. Can you answer him? Can you interpret the answer? Can you do even better than he did?

76. A POCKET MONEY PROBLEM

Jessica's friends Anna, Belinda and Cathryn were comparing their pocket money. Jessica reported that Anna had one third of what Belinda and Cathryn together had and Belinda had one half of what Anna and Cathryn had. Cathryn had some fraction of what Anna and Belinda had, but Jessica didn't hear what Cathryn said. Can you figure out what fraction Cathryn said? What are the least amounts the girls could have had?

77. TAKE YOUR SEATS

I was recently looking through a puzzle book by one of Britain's best known compilers, but I had best not name him as he was an MP! Some of his problems are great classics, and he often gives answers in the classical manner — i.e. he just gives an answer with no indication of how it is obtained nor whether there might be other answers. In some cases, he makes the problem more difficult by leaving out some of the information. The following is an exceptional example of these properties — I have abbreviated it to save space.

A theatre has seats for £3, £4 and £5. All the seats were sold and the theatre took in £7500. How many seats were sold at each of the prices?

Of course, this seems reminiscent of the Hundred Fowls type of problem, which originated in 5th-century China and has been popular ever since. But if this was one of those problems, then the number of seats would have to be given. Even then, there is usually a number of possible solutions. So I glanced at the answer and saw that he asserts that they sold 503 £3 seats, 1494 £4 seats and 3 £5 seats. This gives a round number of 2000 seats in the theatre. Show that knowing the number of seats still leaves a great many more solutions than he has given. And how many possible solutions would there be to the problem as stated — i.e. with the number of seats unknown?

78. A MILLENNIUM CONUNDRUM

When the Millennium was approaching and the Dome and the Jubilee Line were still in a state of chaos, it was clear that we needed to use the extra year available as the next Millennium really didn't start until the beginning of 2001. This has very little to do with our problem except that it involves dates! Reading an 18th-century book, I was inspired to ask how many solutions are there to $19X + 99Y = 1999$ with positive integers X, Y? After doing the whole thing out, I realized that I knew one solution, namely $19 \cdot 100 + 99 \cdot 1 = 1999$ and one easily sees there is only one other solution $19 \cdot 1 + 99 \cdot 20$. So I wondered about using 19 and 99 to make 2000. That is, how many solutions are there of $19X + 99Y = 2000$ with positive integers X, Y?

79. A MAGIC CUBE

A problem in the UK Junior Mathematical Olympiad 1993 considered a cube with the numbers $1, 2, \ldots, 8$ placed at the corners so that the four numbers at the corners of each face always have the same sum. This is a kind of magic cube. Surprisingly, I have run across a form of this problem in *Laugh Magazine* No. 26 which seems to date from the 1950s. The problem asked in the Olympiad was rather simple — show that diagonally opposite edges always have the same sum of values at their ends. I am sure that readers will have no difficulty in finding all possible magic cubes of this type.

Chapter 6

Alphametics

An alphametic is an arithmetical calculation in which some or all the digits are replaced by letters and the object is to deduce what the original digits were. Perhaps the simplest example is $A + A = A$ which has the unique solution $A = 0$. One of the earliest examples is $SEND + MORE = MONEY$ given by Henry Dudeney in *Strand Magazine* of July 1924. I had always assumed this had a unique solution, but I checked with a computer and found 35 solutions. Some time later Don Knuth told me there was only one solution and kindly pointed out some false answers in my list of solutions. I then discovered a typographical error in my program, reran it and confirmed there is just one solution. This shows how easy it is to make a mistake.

Sometimes the letters corresponding to the digits $0, 1, \ldots, 9$ spell a word. The earliest known example of an alphametic is in the *American Agriculturist* of December 1864 where the digits correspond to *PALMERSTON*.

The word 'alphametic' first appears in J. A. H. Hunter's column 'Fun with figures' in the Toronto *Globe & Mail* of 27 October 1955. A later author says "Hunter received a letter from a reader referring to an 'alphametical problem in which letters take the the place of figures'." The names 'cryptarithm', 'arithmetical restoration' and 'skeleton arithmetic' are also used, though a skeleton arithmetic usually has the digits all (or mostly) replaced by a single symbol, such as '?' or '*', as in Problem 80 below.

In May 1931, MINOS [pseudonym of Simon Vatriquant] wrote about such puzzles in the Belgian puzzle magazine *Sphinx* and introduced the word 'cryptarithmie' and stated the basic rules: "A charming cryptarithm should (1) make sense in the given letters as well as the solved digits, (2) involve all the digits, (3) have a unique solution, and (4) be such that it can be broken by logic, without recourse to trial and error." With regard to point (4), one must realize that it is easy to make mistakes in these problems and I have checked results with my computer — but it is also easy to make mistakes in programming, as seen above.

80. SKELETON DIVISION IN THE CUPBOARD

A book dealer recently found the following unsigned puzzle on a postcard in a copy of one of Martin Gardner's books and kindly sent a copy to me. In case you haven't seen one before, the object is to determine all the digits in the following long division diagram, where each '?' denotes a digit. How many possible answers are there?

```
         ??.???
   ??| ????
        ??
        ???
         ??
          ?.?
          ?.?
           ?.??
           ?.??
              ??
              ??
```

81. A PAN-DIGITAL FRACTION

My late Japanese colleague, Nob Yoshigahara, gave me this problem. Use the nine positive digits once each to make the following true.

$$A/BC + D/EF + G/HI = 1.$$

Here the denominator is to be interpreted as a two-digit number, so we interpret 3/24 as 1/8, etc. Nob didn't give me the answer, saying the answer is unique, up to permuting the three fractions. He used it on a marathon program in Japan and said he was surprised when a listener phoned in the answer within a few hours. No one seems to have found a really easy way to find the solution — I certainly gave up and used my computer. Can you find the solution, especially without using a computer?

82. EVEN MORE MIXUPS

In 1993, David Vincent, of Watford, sent in a solution to a problem by one of my co-setters on *The Daily Telegraph*, whose solution was $27-24 = 72/24$. However, Vincent's card was sent on to me by mistake and I thought he was commenting on one of my problems which involved products like $46 \times 96 = 64 \times 69$. My general problem was $ab \times cd = ba \times dc$, so I interpreted Vincent as saying he had found a solution to $ab - cd = ba/cd$. I wondered if there were any more solutions. After a bit of searching, I thought I'd need to let my computer loose on the problem, but then I managed to find the solutions by hand. Can you find all the solutions of this form?

This problem may not be too exciting, but it is a fine example of how problems get created — one person has a pattern arising in one context and another person sees it in a different context and extends it in directions the first person would not have considered.

83. TWO WRONGS CAN MAKE A RIGHT

In reading old puzzle books, one is sometimes amazed how lazy writers seem to be. Then one tries to do the problem and finds it takes a long time. A 1971 book asks for solutions to the cryptarithm: $WRONG + WRONG = RIGHT$. As usual, the different letters stand for different digits and when the numerical values are substituted, the resulting arithmetic must be correct. The solution simply says: "Two possible solutions are" given by $WRONG =$ 24765 or 24153. I thought "Surely I can find all solutions" and indeed I found several times as many solutions. However, it's very easy to make mistakes in doing this kind of work, so I then programmed my computer to do it and it found several more solutions that I had overlooked. So if you try this, be prepared for several pages of careful work. If you want to limit your work, try finding the solutions where $O = 0$ or $I = 1$.

84. THREE LEFTS MAKE A RIGHT

The previous problem asked to show that two wrongs can make a right when interpreted as a cryptarithm. In England and other countries where one drives on the left, it is not easy to make a right turn so we have that "Two wrongs don't make a right, but three lefts do make a right". Find solutions of the cryptarithm: $LEFT + LEFT + LEFT = RIGHT$.

In the US and countries where one drives on the right, the phrase is: "Three rights make a left", but this doesn't give a cryptarithmic problem.

Warning: because this cryptarithm has very little structure, there are a number of solutions and it will take a lot of work to find them by hand. If you don't want to do that much work, just find one solution or assume that no letter represents the digit 0. This reduces the problem considerably. In general, we prefer this type of problem to have very few solutions, preferably a unique solution, which can be determined logically rather than by tedious checking of cases, but I like the meaning of the statements involved in these two problems.

85. OXO CUBE

On the Puzzle Panel program of 16 July 1998, I. M. Berry of Coventry proposed the question "Nothing squared is a cube — explain." My immediate approach was to view it as a cryptarithm or alphametic — i.e. consider $NOTHING \times NOTHING$ is a cube, where different letters represent different digits. However, the answer was *OXO*! [In Britain, this is a well-known brand of bouillon cube.] But, when I got home, I looked for solutions of the alphametic problem. I found there are a few answers, but I did use my computer. Perhaps you can find them — let me know if you find an easy approach.

86. STIR TWO WHEAT

In a book called *World's Trickiest Puzzles*, one expects some interesting puzzles, but this one was a bit trickier than expected. This is based on diner slang "Stir two, wheat" which denotes "Two scrambled eggs with wheat toast". The problem is to solve the alphametic or cryptarithm: $STIR + TWO = WHEAT$, where the letters represent digits. The answer given is $9754 + 713 = 10467$. But this has $R = E = 4$. Such repeated values are normally prohibited in problems of this type, but the statement of the problem actually doesn't prohibit it. Allowing repeated values makes the problem easier in that the conditions are less restrictive, but it is also harder in that it has many more possibilities and usually many more solutions.

Show there are no solutions without repeated values, unless we break another normal rule and permit $W = 0$. How many solutions are there with $W = 0$?

Suppose we do allow repeated values and leading zeroes. The simplest solution here is $0000 + 000 = 00000$. How many solutions are there? And how many have no leading zeroes? This seems to be the version intended in the book. In this case, we must use at least the digits 0, 1, 9 — are there then any solutions using only these digits?

87. FLY FOR YOUR LIFE

World's Trickiest Puzzles gives another alphametic or cryptarithm:

$$FLY + FOR + YOUR = LIFE.$$

With the normal restrictions that distinct letters correspond to distinct digits and there are no leading zeroes — which the author does not state — there are 94 solutions. But the author hints that $O = 0$ and $I = 1$, though what he means is not that these are necessarily true, but he is asking you to assume these are true. Then there is a unique solution which you will have no trouble finding. Elsewhere this version of the problem is attributed to Henry Dudeney. However, if we do allow repeated values, as the author has done in other problems, there are more solutions. Assuming there are no leading zeroes, how many solutions are there? [And how many are there if we permit leading zeroes?]

88. TWO ODDS MAKE AN EVEN

Doodling the other day, I decided to see whether two odds made an even in the alphametic sense. That is, can we have $ODD + ODD = EVEN$? Here we assume the normal rules that the letters represent digits, with distinct letters being distinct digits and leading zeroes are not permitted.

Having done that, I decided to see whether three *ODD*s could make an *EVEN*.

Clearly this can continue and I have used a computer to find all possible answers, but each case takes a little time to do by hand, so this is enough for one problem.

89. ODDS AND EVENS — 2

Having shown that two *ODD*s can make an *EVEN* and that three *ODD*s cannot make an *EVEN*, I suggest you see if this pattern holds for four and five *ODD*s.

90. ODDS AND EVENS — 3

Having looked at a few numbers of *ODD*s making an *EVEN*, perhaps you would like to consider six, seven, eight and nine *ODD*s making an *EVEN*.

If you have a computer, or a lot of time, you could look at even more *ODD*s making an *EVEN*, say $k * ODD = EVEN$, where k is an integer whose value is not being compared with the digits in the rest of the expression. There are a moderate number of solutions. However, there are solutions of $K * ODD$ or $KL * ODD = EVEN$ or $KK * ODD = EVEN$, where distinct letters represent distinct digits. We have already seen several examples of this type where the multiplier is a single digit. Can you find an example with a two-digit multiplier *KL*? There are only a few examples with two-digit multipliers *KK*. Can you find them all?

91. THREE SQUARES AGAIN

In another problem one of the solutions involved 49, and I noted that we have two squares 4, 9 such that the number formed by concatenating them, i.e. 49, is also a square. I wondered if there were more examples and wrote a program to search for them. This soon showed that there are infinitely many examples. Can you see why? I call these the easy examples. My program started with the large square and tried splitting it into two squares. The program produced some improper solutions with leading zeroes, e.g. 9025 is formed from 9 and 025. Eliminating the easy and the improper examples leaves only a few examples where the concatenation is less than 10000. I don't see any easy way to find these, but it shouldn't take you too long to determine them.

92. A FOOTBALL GAME PROBLEM

The on-line maths club NRICH (http://nrich.maths.org) gave the following problem in their puzzles for Science Week 1999: $FOOT + BALL = GAME$.

How many solutions are there? This is described as 'very puzzling???' and no answer is given.

Though not stated, the usual rules apply: each letter stands for a digit with different letters being different digits. It turns out that a complete solution is tedious to do by hand, so what you should look for is a description of how to find solutions. To get all the solutions, you will probably want to use a computer, perhaps even before trying to describe the solutions.

93. SOME PAN-DIGITAL SUMS OF FRACTIONS

A bit earlier, I gave a problem of finding a distribution of the nine positive digits as the letters A, B,..., I to make the following sum of fractions correct.

$$A/BC + D/EF + G/HI = 1.$$

(The denominator is interpreted as a two-digit number, so we view 3/24 as 1/8, etc.)

Mike Bennett of Wokingham looked at the expression $A/BC + D/EF + G/HI$ and determined the minimum and maximum values it can take on — again assuming the letters represent the nine positive digits. A little trial and error will show that relatively few forms have to be examined, but it is easy to fail to look at all of them. Mike Bennett then used a computer to check and determined that he had found one value correctly but not the other. How well can you do?

94. MORE PAN-DIGITAL FRACTIONS

Earlier, I gave a problem due to Nob Yoshigahara: Use the nine positive digits once each to make the following true.

$$A/BC + D/EF + G/HI = 1.$$

Here the denominator is to be interpreted as a two-digit number, so we interpret 3/24 as 1/8, etc. The answer is unique, but not easy to find by hand.

Nob visited and told me that one of his correspondents had found that there is a unique solution to the same problem when BC means $B * C$, the product of B and C.

That is: Use the nine positive digits once each to make the following true.

$$A/(B * C) + D/(E * F) + G/(H * I) = 1.$$

To make this unique, we make the obvious restrictions that

$$A < D < G, \quad B < C, \quad E < F, \quad H < I.$$

95. EVEN MORE PAN-DIGITAL FRACTIONS

On his visit, Nob Yoshigahara said even more. Not only does the problem: Use the nine positive digits once each to make the following true

$$A/BC + D/EF + G/HI = 1$$

still have a unique solution when BC means $B * C$, the product of B and C, but it also has an essentially unique solution when BC is replaced by $B + C$, the sum of B and C.

That is: Use the nine positive digits once each to make the following true.

$$A/(B + C) + D/(E + F) + G/(H + I) = 1.$$

To make this essentially unique, we again make the obvious restrictions that

$$A < D < G, \quad B < C, \quad E < F, \quad H < I.$$

96. SEVEN DAYS MAKE ONE WEEK

I found a nice puzzle book from 1975 [Alan Ward; *Simple Science Puzzles*; [From *Science Activities*, US, 1970–1973]; Batsford, 1975, pp. 21 & 22.] which asked for solutions to $7 * DAYS = WEEK$, where the usual conditions apply — distinct letters denote distinct integers. The author gave the values of E, and made an implicit assumption that led to a unique solution. However, I can't even see a simple argument to find this, so it seems easiest to write a program. How many solutions do you find? And what did the previous author assume?

Chapter 7

Sequence Puzzles

Sequence puzzles ask you to find the reason for a particular sequence and/or to find some missing terms. There is a standard technique for simple sequences — take the differences of consecutive terms and see if you can find a pattern. If not, take the differences of the differences, but such sequences are too straightforward to be considered problems. Instead, one considers sequences such as 8, 4, 5, 9, 1, 6, 7, 3, 2 or 16, 06, 68, 88, ?, 98 where a different point of view is required. In the first, one has to think of the numbers as words and then one sees they are in alphabetical order. In the second, one has to turn the page around to see that the numbers are then 86, ?, 88, 89, 90, 91 — I recently received the latter as an Internet message which said it was Hong Kong Elementary School Admissions Test Question #21. Good luck with the following.

97. A SNEAKY SEQUENCE

Find the next two terms in the following sequence:

10, 11, 12, 13, 14, 15, 16, 17, 20, 22, 24, 31, 100, 121.

98. WHATEVER NEXT?

Here is one of those dreadful sequence problems. What comes next in the following sequences?

A. 2, 4, 6, 30, 32, 34, 36, 40, 42, 44, 46, 50, 52, 54, 56, 60, 62, 64, 66.
B. 1, 4, 5, 6, 7, 9, 11.

99. SENT FOR A TEASER!

I don't know why, but most of us get a perverse pleasure out of making up sequence puzzles. While invigilating an exam, I was reading a puzzle book and the following pattern occurred: SENT. I then realized that this can be continued as: SENTTTTTTTTTTTTTTTTTTTT.

What is the next letter?

100. A PERVERSE SEQUENCE

Jessica's cousin Werner works for British Aerospace and has a rather perverted sense of humor. He recently sent her the following postcard:

Dear Jessica,

The following is an important sequence here. TNESSFFTTO. However, I've left out the last letter. Can you find it? Actually there are several possibilities. How many can you find?

Love from Werner

101. A PENROSE SEQUENCE

Sir Roger Penrose, OM, recently Rouse Ball Professor of Mathematics at Oxford and Gresham Professor of Geometry in London and author of *The Emperor's New Mind* and *Shadows of the Mind*, is also distinguished for several contributions to recreational mathematics. In 1958, he created the Penrose 'Impossible Triangle' which M. C. Escher used as the basis of 'Waterfall', one of the iconic images of the 20th century. In 1972, he invented the 'Penrose Pieces' which tile the plane, but only non-periodically, and which led the concept of quasicrystals and the discovery of a new type of solids — for which the 2011 Nobel Prize in Chemistry was awarded. Some years ago, he showed me the following sequence and kindly allowed me to publish it.

$$35, 45, 60, P, 120, 180, 280, 450, 744, 1260.$$

What is the Penrose number P?

This problem really has two parts. The first is to understand the rule of the sequence and this will allow you to extend it in both directions. I might say that this is a proper mathematical sequence — no funny business with spelling of the numbers, etc. The second part of the problem is to determine what P is and this requires some mathematics which ought to be covered at British A-Level but might not occur until a first-year university course in calculus. So if you don't know what to do, ask a friend who's done some university mathematics (or a course that uses mathematics, such as physics or engineering).

102. MORE SEQUENCES

Here are some more of those dreadful sequence problems.

A. *E, O, E, R, E, X, N, T*, ?, ?
B. 3, 9, 1, 5, 7, 2, 4, 8, 6 — what is the rule?
C. 7, 9, ?, 2, 1, ?, 10, 9, 11

103. EVEN MORE SEQUENCES

Some more sequence problems.

A. 4, 8, 12, 2, 1, 7, 6, 3, 5, 11, 10, 9. What is the pattern?

B. What happens: at 13 in Greece; at 16 in Spain; at 17 in France and Italy; at 21 in Britain, America and Russia; and never in Arabia, Germany, Israel and Norway?

C. 2, 3, 4, 5, 6, 8, 12, 30, 32, 33, 34, 35, 36, 38, ...
What is the last number in this sequence?

D. 1, 2, 3, 4, 7, 10, 11, 12, 14, 17, 20, 21, 22, 23, 24, 27, ...
What is the last number in this sequence?

E. 3, 5, 6, 7, 8, 9, 10, 11, 12, 13, 15, 16, 17, 18, 19, 20, 23, 25, 26, 27, 28, 29, ...
What is the last number in this sequence?

F. 1, 4, 3, 11, 15, 13, 17, ...

104. EVEN MORE SEQUENCES — 2

In an earlier problem, I asked you to identify the following sequence and tell what the last number in it was.

2, 3, 4, 5, 6, 8, 12, 30, 32, 33, 34, 35, 36, 38, ...

How many numbers are there in this sequence?

105. EVEN MORE SEQUENCES — 3

In an earlier problem, I asked you to identify the following sequence and tell what the last number in it was.

3, 5, 6, 7, 8, 9, 10, 11, 12, 13, 15, 16, 17, 18,
19, 20, 23, 25, 26, 27, 28, 29, ...

How many numbers are there in this sequence?

106. EVEN MORE SEQUENCES — 4

In an earlier problem, I asked you to identify the following sequence and tell what the last number in it was.

1, 2, 3, 4, 7, 10, 11, 12, 14, 17, 20, 21, 22, 23, 24, 27, ...

How many numbers are there in this sequence?

107. AN ELEMENTARY PROBLEM

What is the reason for the following set of letters?

A, B, C, F, H, I, K, N, O, P, S, U, V, W, Y.

HINT: After first using this problem, I learned that A is no longer in this set!

Chapter 8

Logic Puzzles

108. STRANGE RELATIONSHIPS IN MUCH PUZZLING

My village of Much Puzzling has a baker, a brewer and a butcher, like most villages. The other day I was talking to the baker's wife and she remarked that these three jobs were held by a Mr. Baker, a Mr. Brewer and a Mr. Butcher, but of course, no man held the job corresponding to his surname.

"But everyone knows that, even a newcomer like myself!" I responded.

But she continued: "But I'll bet you don't know what Mrs. Brewer told me just the other day. You see, each of the men married the sister of one of the other men. And no man married a girl of the same name as his occupation!"

"No, I hadn't known that. That's quite remarkable."

What was the butcher's wife's maiden name?

109. A HOLLYWOOD MURDER

Inspector Lestrade enters the dining room. The body lies in the middle of the room. Four other people were in the room when the lights went out. Each of them makes a statement.

Alice :	I know who killed her.
Benny :	I killed her.
Carol :	Benny killed her.
Donald :	It wasn't Benny or Carol.

Background inquiries reveal that all the suspects are totally untrustworthy, so Lestrade correctly perceives that they are all lying.

Whodunit?

But that's a bit easy. Suppose instead that Lestrade correctly perceives that just one suspect is lying — then whodunit?

110. THE DERANGED SECRETARY

My secretary, Ms. Flubbit, has done it again. She has taken all of today's letters and put each one into a wrong envelope. So now I have to open up all the envelopes to find out what is in each one. But surely, when I have opened all but one of the envelopes, I can deduce what is in the last one. Can I do better? That is, how many envelopes do I have to open before I can deduce what is in each of the remaining envelopes?

111. IN THE PAWNSHOP

My uncle, George King, runs a pawnshop in Castle Square. He says he gets bored in the days, but at night there's a queen who keeps trying to rook him and his mate with bad checks. Anyhow, he says the following happened to him. A student needed $150 to take his girl out to a big dance. All he had was a $100 bill. He pawned the bill for $75 with Uncle George, then he sold the ticket to a friend for $75 — "Just what it's worth," as he said. So he now had the $150 he needed for the dance and was happy. Did anybody lose? Who and how much?

112. A TAXING ROAD PROBLEM

There are four farmers, named Able, Baker, Charlie and Dog, who live up a dirt road leading off State Highway 1. They are located 1, 2, 3, 4 miles, respectively, from the highway. The County offers to pave the road up to Dog's place, if the farmers will pay the cost of it, which the County reckons to be $4800. So the boys get together at Jimmy's Bar and Dog proposes that they all pay a quarter of the cost. The others know that he's a sly old Dog and so they're a bit skeptical. If you are Able, propose a reasonable allocation of the costs.

113. TRUE OR FALSE?

Among the Islets of Langerhans are two known as Truth and Falsehood because their inhabitants always speak the truth and always lie respectively. Despite the closeness of the islands and their isolation from other islands, the inhabitants, naturally called Truthtellers and Liars, are antagonistic to the point of killing any stranger suspected of being from the other island — and they are not really aware of any other islands, despite having learned English from a *Bounty* mutineer. If a stranger is found, he is immediately asked the customary questions: "Which island do you come from? Which island is this?" If he gives satisfactory replies, custom now allows the stranger one question. While sailing around the world single-handed, you pass near these islands and are washed overboard, fortunately after having read about these local customs. You are washed up on the shore of one of the islands, though you do not know which as you cannot see any distinguishing features. A party of inhabitants approaches and asks the customary questions. By now, you have figured out how to respond to these and so you get to ask a question. (What responses should you give?) In order to prevent arousing further suspicions, it is vital to determine which island you are on. How do you do so in one question?

114. HOLMES VERSUS LESTRADE

"It's certainly one of these three servants who has done in their master, Mr. Holmes," pronounced Inspector Lestrade.

"I quite agree," rejoined Holmes, "despite their otherwise unimpeachable histories, it must be Xenia, Yolanda or Zelda."

"Yes sir. It seems quite clear that one of them must be quite cunningly insane — she'll do anything to confuse us."

"She might even tell the truth sometimes."

"My sentiments exactly, sir. But the others know who did it and they will certainly tell the truth. Hence I figure that I can ask each of them one yes or no question in order to determine who is the murderer — just three yes or no questions in all."

"True, but I would only need two such questions."

Can you do as well as Lestrade or as well as Holmes?

115. A STRIKING PROBLEM

My village church clock strikes the hours and makes one stroke on the half-hours. The other night I had trouble sleeping. I woke up and lay there listening to the clock. What is the longest time I could have been awake before I knew what time it was? And what time(s) could it have been then?

116. A HAIRY SITUATION

Looking through old puzzle books often turns up interesting problems. The following appeared in A. Cyril Pearson's *Twentieth Century Standard Puzzle Book* of 1907.

"If the population of Bristol exceeds by two hundred and thirty seven the number of hairs on the head of any one of its inhabitants, how many of them at least, if none of them are bald, must have the same number of hairs on their heads?"

He gives the answer 'at least 474'. Do you agree? What interpretation might Pearson have been using?

117. IN THE DARK

I keep my socks and gloves in my top drawer. There are five pairs of black socks and five pairs of brown socks, as well as five pairs of black gloves and five pairs of brown gloves. The other night, the light bulb burned out in the bedroom when I came in to get some socks and gloves. I can distinguish socks from gloves in the dark, but I couldn't tell black from brown or right from left, so I had to pick up some of each and carry them into the hall to see what color they were. How many items do I have to pick up in order to be sure of having a pair of matching socks and a pair of matching gloves when I go into the hall?

If you can tell me that, then you should realize that I would obviously want the socks and gloves to be of the same color. How many items do I need then?

118. STRANGE RELATIONSHIPS IN LESS PUZZLING

Not far from the village of Much Puzzling is the village of Less Puzzling. As in all country villages, there are a butcher, a baker and a brewer. However, their names were Smith, Jones and Robinson, and their occupations are not relevant here! Strangely, they were all widowers of about the same age with daughters of about the same age. Last spring, there was a grand triple wedding when each of the men married one of the daughters. What is Mr. Smith's father-in-law's father-in-law's wife's maiden name?

119. FERRYING FOUR JEALOUS COUPLES

One of the classic ferry problems, which first appeared in the 9th-century *Propositions for Sharpening Youths* attributed to Alcuin of York, c800, has three jealous husbands with their wives who wish to cross a river using a boat which only holds two persons. Each husband is so jealous that he will not permit his wife to be in the presence of another man without the husband being present. It is not hard to solve this with 11 boat trips or crossings. In 1556, Tartaglia said it was possible for four couples to cross, but he got three women to the other side, then sent two of their husbands across and had the third wife bring the boat back. However this would outrage the jealousy of her husband, so it is not permitted by the rules. Can you show that there is no way to ferry four jealous couples?

In 1879, a M. Cadet De Fontenay suggested that four (or more) couples could be transported if there was an island in the river and gave a solution for four couples with 26 crossings, but he didn't let the boat travel from one side of the river to the other directly. Can you find such a solution and show it is minimal? In 1917, Henry Dudeney found a solution in 17 crossings, where he allowed the boat to go from bank to bank directly and he said this was minimal. Can you do as well as Dudeney — or better?

[This is actually three questions.]

120. FERRYING A FAMILY

A father is traveling with his N children, whom we will call $1, 2, \ldots, N$, in decreasing order of age. Child i is one year older than child $i + 1$. As in all large families, a child gets on perfectly well with his brothers and sisters except for the ones just one year older or younger — as soon as two children a year apart are away from the father, they start to fight. The family comes to a river and they find a boat which can only hold two people. Only the father and the eldest (number one) child can row. How can they get across while preserving domestic tranquillity — and, of course, one wants to have the minimum number of crossings to avoid overwork.

121. HAPPIER FAMILIES IN MUCH PUZZLING

You may recall that in Less Puzzling there were three widowers with marriageable daughters who all got married. In Much Puzzling, we had four handsome and charming widows with marriageable sons: the Archers, the Bakers, the Cobblers, the Dyers. All of them got married at a great quadruple wedding a while ago. Needless to say, this made it quite complicated talking about family relationships, especially since nearly everyone in the village was related to one of the families and hence is now related to nearly everyone. The Much Puzzlers soon developed a short-hand system, e.g. one's step-father's step-father is called one's 'second step-father', with 'second step-son' being analogous. The other day, I was talking to Mr. Archer. Being an outsider, I have trouble remembering who is married to whom and the Much Puzzlers always are teasing me. I asked Archer a question about the new Mrs. Baker which confused him. "I'll have to consult with my fourth step-son about that", he said. Mr. Cobbler was nearby and burst out laughing — "You old so and so, Archer." When I asked what was so funny, he just said "You just think about it." I protested that I didn't remember who was married to whom and he repeated his statement and went off chortling. It took me some time to see why Cobbler was so amused. Can you explain why — and who was Archer's fourth step-son?

122. FOUR GREAT-GRANDPARENTS

Some years ago, I read a description of some English aristocrat which said that he only had four grandparents while most people had eight. This was later corrected — he had only four great-grandparents while most people had eight. Of course this puzzled me for a bit and I worked out a reasonable way this could have happened. Recently, however, I have realized that there is another way it could have happened, though I think the second method is slightly less likely. Can you find both ways?

Chapter 9

Geometrical Puzzles

123. HOME IS THE HUNTER

Hiawatha, the mighty hunter, has wandered far in search of game. One morning he has breakfast at his camp. He gets up and heads north. After going ten miles in a straight line, he stops for lunch. He eats hurriedly, then gets up and again heads north. After going ten miles in a straight line, he finds himself back at his morning's camp — honest Injun! Where on earth is he?

124. THE STABLE TABLE FABLE

My cousin Mabel gave me a circular table for my patio and it wobbles. I have checked and I find that the table is not at fault — it has no wobble when it stands on a flat floor and the feet of the legs form a perfect square. Show that when I place the table on my patio, I am able to rotate it so Mabel's table is stable.

125. MATCHSTICK TETROMINO QUADRISECTION

Consider an array of 4 by 4 square cells with the outer boundary drawn, as shown in the first figure. Make the figure so that it is four matchsticks long on each side. We wish to divide this into 4 connected regions of area four (i.e. tetrominoes) by placing matchsticks on the figure. For example, the second figure gives one obvious way to do it, which uses 12 matchsticks. How many matchsticks may be needed?

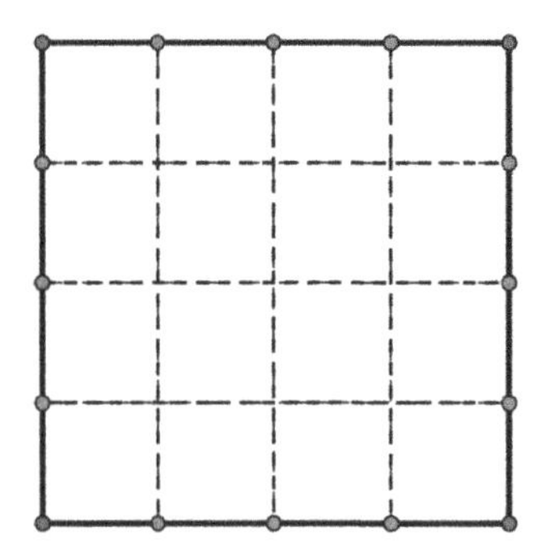

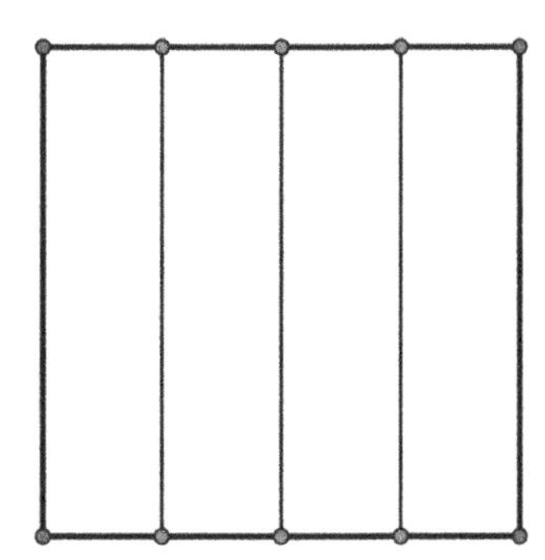

126. OFF THE RAILS

When browsing through puzzle books, I often come across familiar problems, but sometimes the answer is unfamiliar. Usually I find this is due to my faulty memory or to the problem being changed in some subtle way. However, here is a problem which is genuinely the same in two books, but the two books give much different answers.

If a railroad rail a mile long is raised 200 feet in the center, how much closer would it bring the two ends?

Robert Ripley's *Mammoth Believe It or Not* (London, 1956) says "less than 6 inches". Jonathan Always' *More Puzzles to Puzzle You* (London, 1967) says "approximately 15 feet".

Which (if either) is right? What value must 200 feet be replaced by to make the incorrect one(s) right?

127. LEANING TOWERS OF NEW YORK?

The two towers of the Verrazano Narrows Bridge are 1 mile apart at sea level and 1 mile plus 1 5/8 inches apart at the top. How high are the tops above sea level?

[In case you want to know, the radius of the earth can be assumed to be 4000 miles.]

128. SQUARE CUTTING

Take a square sheet of paper, perhaps a paper napkin. Using one straight cut with a scissors, divide it into four equal squares. Once you've done that, work out how to divide it into four equal triangles.

129. A SIMPLER DISSECTION

The *Morley Adams Puzzle Book* of 1939 asks to cut a square into four pieces which can be assembled into the following shape.

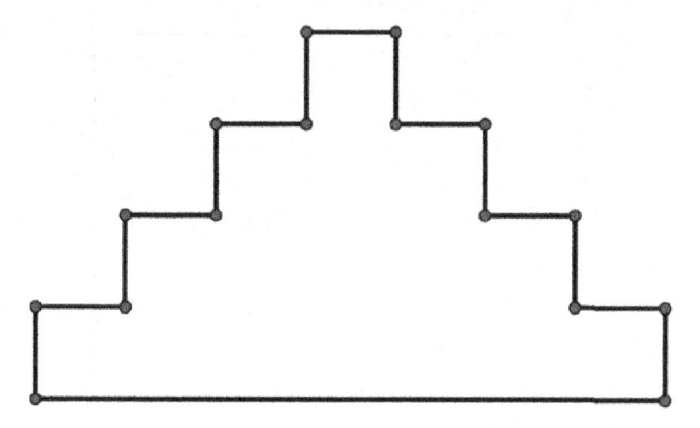

Show how to do it with only two pieces! Adams' solution shows that he meant four congruent pieces. Can you do what he meant?

130. CUTTING THE CHRISTMAS CAKE CORRECTLY

Jessica and four friends wanted to share a small square Christmas cake. This wouldn't be a problem except that Christmas cakes have frosting on the sides as well as on the top and everybody liked the frosting as well as the cake. Making four parallel cuts would give two pieces with lots of frosting and three pieces without so much. They began to argue so loudly that I had to come down and help out. How could I cut the Christmas cake correctly?

131. RETRIEVE THE BOAT

Jessica and I were watching a man running his radio-controlled boat on the pond on Clapham Common. It ran out of fuel and came to a stop in the center of the pond. No one felt like getting wet. There were various sticks and lots of light rope, but the boat was so far out that we couldn't reach it with the sticks, not even tied together, and no one had enough strength or skill to throw the rope out to it, though there was more than enough rope to reach the boat and back again. It wasn't drifting to shore, since there was no wind, nor any sign of wind in the offing. Besides it was getting dark. The pond couldn't be drained. How could we save the boat?

132. BRIDGING THE MOAT

An old problem involves a square moat about a square central island, as shown in the diagram.

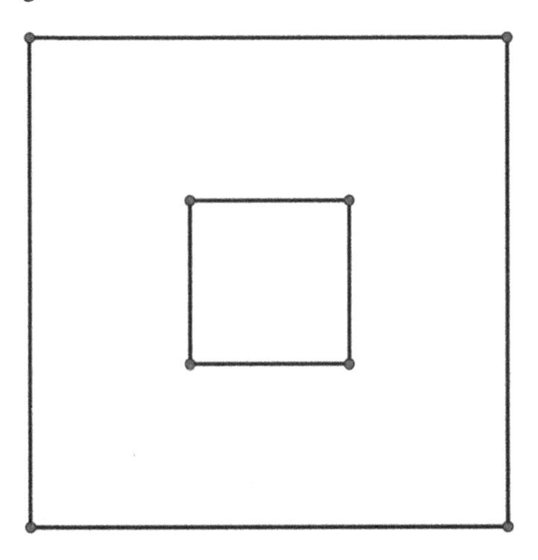

The gap between the island and the shore is 1 unit. We want to build a foot bridge to the island across the moat, which has vertical walls on both sides. There is a supply of planks of length L. Unfortunately, L is just a few percent short of 1 so that a board will not reach directly across the moat. How can you bridge the gap with two boards? How large must L be to do this?

Now suppose the moat is circular with a circular central island. The width of the moat is still 1, but now the problems depend on the radius, r, of the island. For $r = 0$ and $r = 1$, what length of boards is needed to bridge the moat with two boards?

[Extra Credit: Show that if we have enough boards of positive length L, we can bridge the moat.]

133. FIND THE CENTER

Take a pad of paper and a pencil. Trace a largish circular object onto the top sheet of paper. So you now have a circle, but you don't know its center. How can you find the center, using just the pad and pencil?

134. A FALSE CUT

A 1929 puzzle book shows a 5 by 5 piece of plain fabric from which a young seamstress has cut out a Greek cross of 5 unit squares as shown.

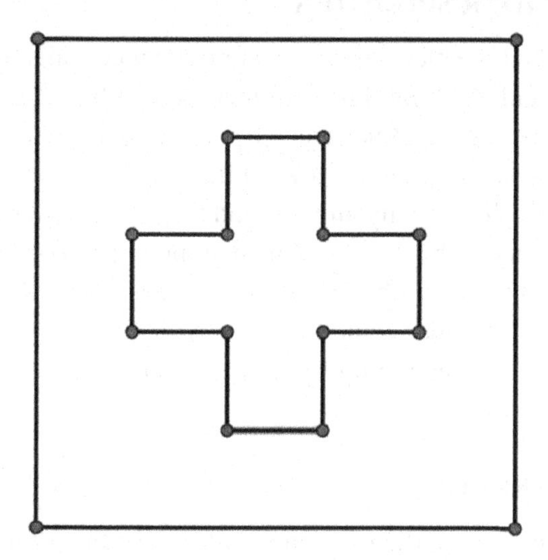

Her mother wishes to reform the piece into a square by cutting it into just two pieces. The answer given was to cut it as shown.

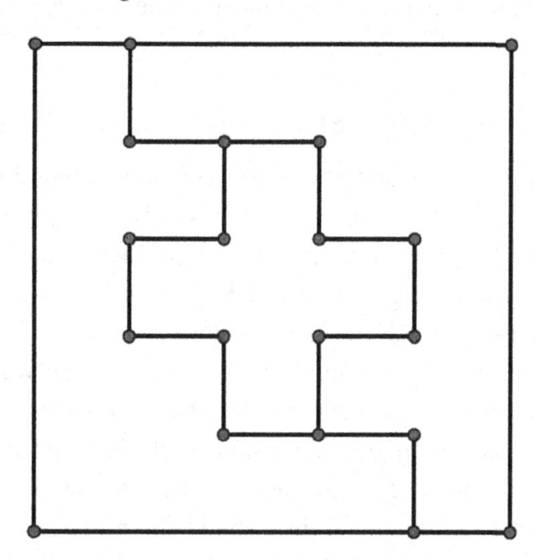

What's wrong with this solution? And what must be wrong with the problem?

135. THE SPIDER SPIED HER

In our garden shed is a piece of thin clear plastic pipe which I recall is 6″ in circumference and 30″ long. I noticed that a spider has taken up residence inside it, 6″ from one end. Whenever a fly alights on the outside of the pipe, the spider immediately marches toward her along the shortest route. The other day, I observed that a fly alighted and the spider started to move, then stopped and changed direction and then stopped and changed again and again. I realized that the problem was that there were two shortest routes to the fly and the poor spider couldn't decide which way to go. Where could the fly have been? Could the fly be anywhere so the spider has even more shortest routes?

136. SNOOKERED!

You are all familiar with the triangular pattern that the fifteen pool (or red snooker) balls are placed in. This is done with a triangular rack. You can buy moulded plastic racks with rounded corners, but simple cheap racks are made from thin pieces of wood cut on a bevel and nailed together. If the balls have radius r and the wood has thickness t, what are the lengths of the inside and outside of a cheap rack?

137. GOAT AND COMPASSES

Jessica was visiting her friend Bill in the village of Much Puzzling. Bill's family keeps a goat and, as usual, a goat has to be tethered to keep it from wandering away. There is a rectangular shed in the middle of a large field and Bill's father wants the goat to graze the grass around the shed so he has adjusted the rope so it's half the circumference of the shed. "That way the goat can get to all the grass around the shed — half one way and half the other way. Now you two go down and put the stake by the shed, but mind you do it so the goat can graze the most ground." Bill and Jessica went off mumbling. "Does it make any difference where the stake goes?" "I'm not sure. By symmetry, I think it ought to be in the middle." "Maybe, but in the middle of the long side or of the short side?" "Or maybe it should be at a corner?" "Hmmm.... but which corner?" By the time they got to the shed, they were so confused they had to come back and get out pencils and paper and compasses. They drew a lot of pictures and did a lot of scribbling before they announced that it would be much easier to remove the shed! Can you help them out?

138. SQUARE BASHING

Some years ago, I found the following in a Russian problem book. Given four points in the plane, find (i.e. actually construct) a square whose four sides (possibly extended) go through the four given points. About the same time, my colleague Ivan Moscovich told me he had heard the same problem from someone who said he had been trying to solve it for many years — my feeble memory recalls he said 20 years! I mentioned the problem to another colleague, Oliver Pretzel, who produced a fairly complex solution. But the Russian solution can be presented in a short paragraph. Can you find a short solution?

139. SHADES OF EUCLID!

Jessica and Bill were walking down the street one evening and when they passed under a street lamp, they jumped on the shadows of their heads. After doing this a few times Jessica pointed out that they could only catch the heads of their shadows, or the shadows of their heads, right under the lamp, "because the shadow ran away from them pretty fast and it goes faster and faster as we get further away from the lamp post." Bill stopped and starting drawing triangles on the pavement and said "Wait a second — we should be able to work out how fast it goes." After a minute he gave up, saying it was too confusing, but when they got to our house, he asked me if Jessica was right or not — and how fast is the shadow's head going?

140. MATCHSTICK SQUARES

It is easy to make a square with four matches and just as easy to make two squares with seven matches, but can you make two squares with only six matches? The squares must be of equal size, one cannot bend or break matches and a match is not allowed to cross another match.

141. CUTTING UP A SQUARE AGAIN

In the past, we have had some tricky cutting up problems. Here is a new one which I found in some material by Douglas Engel, a well-known puzzler from Colorado.

Dissect a square piece of paper into three identical triangles with a single cut.

Clearly some trickery is involved, but no more than my readers are used to!

142. CUTTING UP A TRIANGLE

Douglas Engel also gives the following. Cut an equilateral triangle with three cuts to produce three identical equilateral triangles.

143. MAKE THREE SQUARES

Most of you will know a number of matchstick puzzles, but here is one that I made up.

Use nine matches to make three squares. As usual, matches may not be bent or broken and are all the same length. Use all the matches.

HINT: There are several answers, some more reasonable than others.

Chapter 10

Geographic Problems

Because we live on an approximately spherical earth, there are a number of unusual situations that can occur which provide problems directly or via the related topic of time-keeping. Here are some.

144. TIME FLIES

If one flies from New York to Los Angeles at about 700 mph, then one keeps up with the earth's rotation and arrives at the same time that one leaves. If one goes faster, say in the late Concorde, then one arrives in Los Angeles at an earlier hour than one departs from New York. Suppose we continue this all the way around the earth. Then it seems that we arrive back in New York before we start! What's wrong with this?

145. SAM'S ESTATE

Slick Sam, the local estate agent, was trying to sell Simple Simon some land. "See here on the map. It's the tidy triangular tract with corners A, B, C."

"But how big is it?", asked Simon.

"Well, I've had it surveyed, but my surveyors are a bit odd, not to mention old-fashioned. The first one sent in the distance from A to B, but he went around via C. He said it was 1 3/4 furlongs. My second surveyor reported that the distance from A to C via B was 27 1/2 chains. Obviously I had to send out a third surveyor to sort out the confusion, but he's just rung up and said that the distance from B to C via A is 100 rods. Land around here is selling at £10, 000 per acre, so I'm sure you'll agree that this is a real bargain at £10, 000."

Can you help Simon decide whether to buy or not?

146. FLAT EARTH?

Charles Fort was an eccentric collector of curiosities, preferably unexplained ones. In his *New Lands*, he reports that in 1870, one John Hampden attempted to prove the earth was flat by surveying a six-mile straight stretch of the Bedford Canal. Fort asserts that the far end of the canal "should be depressed 288 inches", which he explains as 8 times 6 squared. Is this right? (Assume the radius of the earth is 4000 miles.)

147. AN ALL-DAY SUNRISE

Jessica was listening to her uncle, A. Tall-Storey, who was telling some of his fabulous adventures, such as visiting the Islets of Langerhans or exploring the Alimentary Canal.

"One of the strangest things that I ever saw," he said, "was a sunrise that lasted all day."

"Wait a second," said Jessica, "Say that again."

"Well, I suppose I'd better change that to a sunrise that lasted 24 hours. The flying crew and I got up early and as the sun rose, we headed west and flew at a steady pace of 400 miles per hour with the rising sun behind us. I was rather amazed to find that the sun stayed precisely at sunrise behind us. After 24 hours we were back at our airstrip and landed as the sun finished rising."

"Poppycock and hogwash, just like your other stories," said Jessica.

"Don't be so sure, young lady. Sometimes I do tell the truth. In fact I haven't even told you of the other interesting thing that happened when we repeated the process except that we flew to the east."

Could he be telling the truth? If so, where on earth was he? And what happened when he went east?

148. A TALLER STORY

Jessica was very impressed with Uncle Tall-Storey's explanation of how he flew around the world at 400 miles per hour for 24 hours with the sun always at sunrise behind him. After several days of drawing pictures, she came back and said: "If you had done your circumnavigation twice as fast, you would have seen even more remarkable sights." He replied that his plane wouldn't go any faster and Jessica said, "OK, but you would have seen the same things at 400 miles per hour if you did your circumnavigation about half as far from the pole." Uncle Tall-Storey's geometry never was very good, so he couldn't figure out what he would have seen. Can you help him out and tell him what he would have seen and where he could have seen the same effect at 400 miles per hour?

149. A POINT WITH A VIEW

An 1849 algebra book gives the following question which has a surprisingly simple answer.

How high above the earth must you be in order to see 1/3 of the earth's surface?

Obviously, one assumes the earth is spherical, but you may need to know (or recall) a truly remarkable result of Archimedes. The amount of area of a spherical surface enclosed between two parallel planes which cut the sphere (i.e. in a zone) is proportional to the distance between the planes and does not depend on where they are. For example, any two cutting planes separated by a radius of the sphere will enclose half the area of the sphere.

150. HOW TO LOSE A DAY

You may recall that Jessica's Uncle Tall-Storey frequently tells her about the impossible things that have happened to him. Just recently he was visiting and Jessica got completely flummoxed when he said one of the strangest days he had had was the day he didn't have. When asked for more details, he said: "Well, I actually went from the fourth of a month to the sixth in a moment, so I never had the fifth of that month. I would have been very unhappy if the fifth had been my birthday, but the sixth was, so I got to it a day sooner than otherwise. Now you tell me where I was." Jessica puzzled over this for some time and then asked: "Were you at a pole again?" "Nope". Eventually Jessica worked it out, but perhaps you can do it quickly.

Chapter 11

Calendrical Problems

Time-keeping is either large scale (days, years, etc.) or small scale. Larger scale time-keeping is tied up with the earth and hence is somewhat connected to geography, as shown in the last problem and some of the following problems.

151. A GRAVE MISUNDERSTANDING?

Salisbury Cathedral is one of the great glories of English architecture. While wandering through it, I came across this tomb slab in the North Choir Aisle.

This has the following remarkable inscription:

H S E
the body of Tho
the sonn of Tho.
Lambert gent
who was borne
May ye 13 An. Do.
1683 & dyed Feb
19 the same year.

(H S E is an abbreviation for the Latin 'Hic Sepultus Est' — here is buried.) My guidebook describes the tomb and even quotes a bit of doggerel: "Thomas Lambert all should mourn for he died three months before he was born!" Explain this strange situation.

152. WANT A DATE?

Quickly now, what year was it 2,100 years ago?

153. A SHORTER CENTURY

Every reader of these problems already knows that the 21st century didn't start until 1 January 2001 because there wasn't a year 0. But do you know that the 21st century will be shorter than the 20th century. Why?

154. A CALENDAR ODDITY

Jessica and her friend Sarah were looking at the calendar for 1992 and Sarah was bemused by the fact that there were five Saturdays in February 1992. Jessica then wondered when it would happen again, but they couldn't find a calendar that went far enough ahead to tell. Can you help them out?

155. UNLUCKY YEARS

Jessica's history book said that the Depression had really started in 1930. Jessica had just finished her math homework and was looking over her history notes, but she kept doodling with the numbers. After some thought, she exclaimed: "Of course. The digits add up to 13, so it had to be an unlucky year." She showed this to her friend Bill, who said: "Wonderful.

We'd better prepare for the next unlucky year. When is it?" This didn't take them very long, but they noticed there is quite a big gap involved. With all the millennial thinking going on, they wondered if there was anything unusual about this millennium compared to the next one and the last one. They wanted to know if the biggest gaps between unlucky years were bigger or smaller and whether there were more or fewer unlucky years in this millennium or the next one or the last one. They had some trouble doing this, so I helped them do it for each of the first ten millennia (i.e. up to 10,001 AD). Can you manage to do this? Which millennium has the most unlucky days and the fewest? What are the largest and smallest gaps between unlucky years?

156. A LONG MONTH

Which month of the year is the longest? There are two possible answers, but one of them is a quibble. The other depends on where you are!

157. HOW LONG IS A MONTH?

John Conway once posed the question of how many different lengths of month are there? There are two ways to answer this, but the real-time solution has more answers!

158. TWIN TIMES

How can two twins, who are born at the same time and who die together, have seen a different number of days?

159. A LONGER MONTH

A few problems ago, I posed the problem of finding the longest month. September was the longest in terms of letters, but October is the longest in time, being 31 days and one hour, because of the change from summer time to standard time, which normally occurs on the last weekend of October, but may vary with the country.

A reader, John Bolton, of Yeovil, Somerset, sent in an idea which allows for an even longer month, at least if one is willing to change one's point of view. Can you find this?

Chapter 12

Clock Problems

Clock problems date back to the time when minute hands started appearing on clocks in about 1660. By 1700, they were fairly common. The earliest clock problem that I have seen is from 1678.

160. WATCH YOUR CLOCK

I'm sure you have all done some of the many problems involving clock hands. You know the type — when are the hands together or in a line or reversible, etc.? However, I don't seem to have seen many questions involving all three hands. You know: the first, short, hand called the hour hand; the second hand called the minute hand and the third hand called the second hand. Now that you are thoroughly confused, can you tell me whether the three hands can ever make equal angles with one another?

161. THREE CLOCKS

Mathematical puzzles are full of strange situations and exotic phenomena. Translation of such puzzles requires the translator to understand what the puzzle is about — when the translator makes a direct translation, even more peculiar effects may arise. I found a problem with three clocks in a German book by Kirsch and Korn translated as *Number Games*. The three clocks all chimed at midnight together. One clock kept the right time, but one "was always 10 minutes fast" and the other "was always 10 minutes slow". Clearly this is a nonsensical rendering of the problem since the clock which is 10 minutes fast cannot possibly be reading midnight when the correct one read midnight. Clearly it is intended to say that the fast clock gains 10 minutes each day and the slow clock loses 10 minutes per day. When will all three clocks strike 12 together again?

In fact the solution of this puzzle is also confusing. It says that it will be 72 days before the fast and normal clocks strike 12 together again. (10 minutes per day = 1/6 of an hour per day, so it takes $6 \times 12 = 72$ days to gain 12 hours. Note that the normal clock thinks it is midnight 72 days later, but the fast clock thinks it is noon of the next day.) But the solution continues that it will be 60 days before the slow and normal clocks strike 12 together again. It then takes the least common multiple of 72 and 60, which is 360. Can you make sense of this solution?

162. PANDIGITAL TIMES

One way of writing a time of a year is as MM:DD:HH:MM:SS where the pairs of digits represent the month, day, hour, minute and second. My late Japanese colleague Nob Yoshigahara [*Puzzlart*, Tokyo, 1992] asked about pandigital times, i.e. such expressions using all ten digits. However he assumed the first M is 0 and then sought for cases where the remaining digits are all the positive digits. He asked how many such times there are and what are the earliest and latest in the year. I wondered why he had restricted himself to having the first digit be 0 as it seems there might be many more cases. Can you answer his questions and then consider the more general case, determining how many more solutions occur? It is easy to compute the answers, and you might like to do this first, but can you deduce them?

163. A VERY ODD CLOCK

Driving some time ago, I noticed that the clock in my car showed that it was 6:59. A bit later, I observed that it showed 5:01. How could this happen?

164. AN UPSIDE-DOWN TIME

We have a standard digital clock in our bedroom on a small side table with digits displayed on the standard 'seven-bar display' which has three cross bars and four upright bars and uses a 24-hour clock. In rearranging the table, the clock was turned upside down. I noticed that the clock appeared to show the correct time! When might this have been?

165. TIME IN REVERSE

We have a standard digital clock in our bedroom on a small side table with digits displayed on the standard 'seven-bar display' and uses a 24-hour clock. In rearranging the table, the clock was turned to face the mirror. When I looked at it, the mirror image appeared to show the correct time! When might this have been?

Chapter 13

Physical Problems

166. THE SQUASHED FLY

Two locomotives are heading towards one another from 100 miles apart on a straight track. The first is going 60 mph and the second is going 40 mph. A fly starts at the front of the first locomotive and flies to the second and then back to the first, then back to the second, etc. Eventually there is a god-awful crash and our fly is squashed. If the fly can fly 50 mph, how far does he fly before the smash?

167. ROUND AND ROUND AND BACK AGAIN

I recently found the following problem in a 1925 book. It is remarkable in that you are given no numerical values, but are asked to determine a number.

A cyclist and a runner start round a race track. The cyclist goes all the way around and then catches up with the runner. He then instantly turns around and heads back to the starting point where he meets the runner who is just finishing his circuit. Determine the ratio of their speeds.

168. A KNOTTY NAUGHTY NAUTICAL PROBLEM

The nautical distance from Los Angeles to Honolulu is 2260 nautical miles. A ship sets out from Los Angeles and proceeds at 1 knot per hour. How long does it take to get to Honolulu?

169. LUNATIC GRAVITY

The force of gravity at the moon's surface is pretty nearly 1/6 of the force at the earth's surface. If we ignore air resistance, school physics tells us that the position, $s = s(t)$, at time t, of a body falling under the force of gravity, from initial height s_0 and with initial upward velocity v_0 is:

$$s = -\tfrac{1}{2}\, gt^2 + v_0 t + s_0,$$

and that the velocity at this time is:

$$v = -gt + v_0.$$

With a little calculus or trigonometry, one can show that the maximum range of a gun is achieved when one fires it at a 45° angle, which gives the bullet equal vertical and horizontal velocities of $v_0/\sqrt{2}$, and this also applies to javelin throwing.

Ignoring air resistance, which of the following would be true on the moon?

(A) A ball thrown straight up will reach 6 times as high as on earth.

(B) The same ball will take 6 times as long to return to the ground.

(C) A stone dropped down a well will take 6 times as long to hit bottom.

(D) A bullet fired horizontally will travel 6 times as far.

(E) The maximum range of a gun will be 6 times as great — or a javelin thrower will throw 6 times as far.

In all these problems, the projectile is given the same initial velocity as it is on earth. We assume that the projectile remains close enough to the surface that g remains constant.

170. WEIGHT WATCHING

I suppose you all know the old chestnut — "Which is heavier: a pound of feathers or a pound of gold?" Of course the pound of feathers is heavier because it is weighed with an Avoirdupois pound of 16 ounces, while gold is weighed with a Troy pound of 12 ounces. [Incidentally, Troy has nothing to do with the classical city, but derives from the French city of Troyes which had a major market in the middle ages.] So now you think a pound of feathers weighs $4/3 = 1.333\ldots$ times as much as a pound of gold. But this isn't correct yet, because the Avoirdupois ounces and the Troy ounces are not the same. One has to subdivide the Troy ounce into 480 grains in order to get a common unit. The Troy pound has 5760 grains, while the Avoirdupois pound has 7000 grains. So a pound of feathers apparently weighs 7000/5760 = 175/144 = 1.2152777... times as much as a pound of gold. Is it any wonder that most of the world has gone over to the metric system? So let's use a kilogram weight and a balance to weigh out a kilogram of feathers and a kilogram of gold. Now which is heavier?

171. WHEEL TROUBLE

Car enthusiasts will know that the rear axle of a rear-wheel-drive vehicle has a differential gear to allow the wheels to turn at different speeds. This is essential for turning because the outside wheel travels along a circle of larger radius and hence goes further and turns more than the inner wheel. But a railroad car has rigid axles, with the wheels firmly attached at each end. How can a railroad car go around a bend?

172. SCREWED-UP

A cylindrical helix is just a spiral on a cylinder, like an ordinary spring or the thread on a bolt. There are two kinds — a right-handed helix and a left-handed helix. If I turn a left-handed helix over (i.e. end for end), does it become a right-handed helix? Give as simple an explanation as you can.

173. RACING ALONG

A special test circuit has been built for testing fuel economy. It has a mile of flat track, then a mile uphill, then another mile on the flat, then a mile downhill. Then the sequence is repeated, but with two mile stretches, then again with three mile stretches, etc. to a total length of 60 miles, which brings the circuit back to its starting point. It is found that optimum gas mileage occurs when the car does 40 mph on the flat, 30 mph going uphill and 60 mph going downhill. How long does it take to complete the course at these speeds?

174. CAUSE FOR REFLECTION

Jessica was admiring her new torn jeans in the mirror. She was annoyed because the mirror was so short that she couldn't see herself all at once.

Since she is now 5 ft 6 in tall, she demanded that I buy her a mirror at least that tall. I said I would buy her a mirror that was at least tall enough — provided she told me the height of the shortest mirror in which she could see her head and toes all at once, and how far away she must stand in order to do so. She is having trouble solving this. Can you help her out?

[Can you also find the narrowest mirror?]

175. FURTHER CAUSE FOR REFLECTION

At various science centers, one can see a three-way corner mirror. This has three square mirrors, set edge to edge at right angles, as at the corner of a room. Such a corner mirror has the remarkable property that a light beam incident on it is reflected back parallel to itself. These mirrors are used in surveying and an array of them was placed on the moon for use in measuring the distance to the moon.

If the mirrors are big enough, e.g. if they were the floor and two walls of a room, how many images of yourself can you see? When you look at the corner where all three mirrors come together, what do you see?

176. OVERTAKING VERSUS MEETING

Jessica and her friend Stella were doing a one-hour sponsored run for charity around the school's one-mile track. Jessica noted that if they kept up their average speeds, she would overtake Stella twice during the hour and once again at the end of the hour. Stella said: "But if I run the other way, I'll meet you 10 times during the hour and once again at the end." How fast can they run?

177. WELL TRAINED

The late passenger train from London to Newcastle starts at the same time as the early morning freight train starts from Newcastle for London. Since the rails are pretty empty, they each manage to maintain a constant speed over the whole trip. When they pass, the drivers wave to each other. They later compared experiences and the passenger train driver said he got into Newcastle just an hour after they passed. The freight train driver said it took him just four hours to get to London after they passed. What part of its journey had the passenger train completed when it passed the freight train — and how long did each train take for its trip?

178. ANY OLD IRON?

A barge loaded with scrap iron is floating in a water-tight canal lock. Some vandals with more energy than sense throw all the iron overboard into the lock. What happens to the water level in the lock?

This problem has been around for some decades, but I've never seen anyone ask about the barge. Obviously it is higher in the water than before, but is it higher from the bottom of the lock?

179. WATCH THE BOUNCING BALL

Jessica and her friend Jorja were experimenting in the school physics lab. With a good rubber ball, they found that it would bounce back to half the height it was dropped from. They then observed that on the second bounce, it would come to 1/4 of the original height, and so on. Then Jessica said: "If it keeps on bouncing like this, it will bounce an infinite number of times." Jorja was very perplexed by this — "If that's true, it must keep bouncing forever." Neither of them liked this, but they couldn't see what was wrong with the argument. Can you help them out?

180. JUMPING OVER THE MOON

Dick Fosbury jumped 7 feet $4\frac{1}{2}$ inches at the Olympics in Mexico. Assuming gravity on the moon is just one sixth of what it on earth, how high would he jump on the moon? Most people think this is easy — it would be just six times $7' \, 4\frac{1}{2}''$, which would be $44' \, 3''$. Indeed, I recently found this answer in the *Third BBCTV Top of the Form Quiz Book* of 1970. But this is very far from correct. Can you see why and find a better answer?

181. FURTHER REFLECTIONS

Most of us have seen a double mirror in the corner of a room. If the mirrors are at right angles, then one sees a genuinely reversed image in the corner. That is, if one puts up one's right hand, then the hand on one's left in the mirror rises. The image is what would occur if one turned around and stood behind the mirror, so one might think of the raised hand as being the right hand of the image. However, my problem is a bit different. When one looks at the corner, one sees a whole image of one's self — except for some loss due to the joint between the mirrors, which produces a bit of a line down the middle of one's image. What do you see when you close one eye? Now shift to using just the other eye.

182. A CRAFTY WEIGHTLIFT

A weight lifter can lift 500 kg from the floor. He has to lift somewhat more than this. So he arranges a simple pulley system above the weight. This has two pulleys. One is attached to the ceiling beams. A rope passes over it, then down through a lower pulley and then is attached to the center of the upper pulley. The center of the lower pulley is attached to the weight. This is readily seen to give a mechanical advantage of two to one. What is the heaviest weight he can raise with this apparatus?

REFERENCE

Alan Ward; *Simple Science Puzzles*; [From Science Activities, US, 1970–1973]; Batsford, 1975, pp. 25 & 27.

Chapter 14

Combinatorial Problems

Some combinatorial problems are ancient, but it has only been since about 1940 that combinatorics has been recognized as an interesting and useful branch of mathematics. When I was a student in the 1960s, there were only a handful of textbooks in combinatorics. There are now hundreds of books and dozens of journals in the field and it is also recognized as an important part of computer science.

Basically, combinatorics asks 'how many?' How many ways can you throw two (or n) dice (or coins)? How many routes are there through a network? How many ways can you get through a maze? How many ways can you arrange a tournament? There are several levels of answer. In some problems, one just wants to know whether the answer is zero or positive. In problems where the answer is positive, one can ask 'how many' solutions are there — or one can ask how can one find or describe the solutions — this is especially the case when there are infinitely many solutions. We have already seen a number of problems where the number of solutions is wanted and several where all the solutions are given. Here is a selection of combinatorial problems.

183. PATIO PATH PAVING

I have a path up to my patio which is two feet wide and ten feet long. I have bought ten paving slabs which are one foot by two feet. How many different ways can I pave my patio path? For example, if the path were just two feet long, then I could put the two slabs down crossways or longways to get two different pavings.

184. HALF A CUBE

Some years ago, during the Rubik Cube® craze, the German firm of Togu made $2 \times 2 \times 2$ cubes with various color patterns. The simplest one had 4 red cubelets and 4 blue cubelets. The advertising material said that there were 70 different patterns on this cube. This was based on the fact that we can choose 4 of the 8 cells in a $2 \times 2 \times 2$ array in $8!/4!4! = 70$ ways. However, these choices are not really all different. One such choice is to make the bottom half red and the top half blue and this is really the same thing as choosing any one of the six halves as red. That is, there are six different choices which give the same pattern. So how many really different patterns are there?

185. SOLID DOMINOES

Consider a 3×3 chessboard and a supply of dominoes which cover two adjacent cells of the board. Clearly one cannot cover the 9 cells of the board with dominoes, but it is easy to cover the board if we omit the middle cell. Now consider a $3 \times 3 \times 3$ cubical array of cells and a supply of 'solid dominoes' or 'bicubes', i.e. blocks which cover two adjacent cubical cells of the array. Again we cannot expect to cover the 27 cells with solid dominoes, but can we do it if we omit the middle cell of the array?

186. ALL TIED UP!

Harry the Horse and Big Julie were arguing about a three-horse race. Harry said there were three ways it could end — either A or B or C could win. But Julie pointed out that which horse came in second was important, so there were six ways the race could end. Harry seemed happy with this, but a bit later, he said: "What about ties?" Both Harry and Julie tried to work out the number, but they weren't sure they got it right. Can you find the number of ways when ties are allowed? If that's too easy, try to find the number of ways in a four-horse race.

187. DICING AROUND

Jessica was playing Monopoly the other day and couldn't find the dice, so she borrowed some from my rather random collection. She chose dice with spots or pips, rather than any with numbers because she says it is hard to read the numbers when they are sideways or upside down. After a bit, she said: "These dice are different."

"Well, that's clear", I responded, "The bigger one is blue and the smaller one is red."

"No, no, don't be trivial! I mean they're really different!"

"That's possible, the standard arrangement has the opposite faces adding to seven and perhaps one of these was made with a non-standard arrangement."

"No, that's not it. The opposite faces do add up to seven."

"There's another possibility. One die can be the mirror image of the other."

"Oh, I hadn't thought of that. But that's not it. These are different in a different way — some of the numbers are printed in different directions on the die. On this one, the three points toward this corner, while on this one, the three points to that corner."

How many different standard dice can there be?

188. COUNTING NUMBERS

The English are terrible about putting up house numbers — maybe it's because they really don't want any visitors or perhaps they're still preparing for an invasion. When I bought my cottage in Much Puzzling, no previous owner had put up the number, which is 123. So I went down to the hardware shop to get some nice numbers for the gatepost.

The proprietor, Mr. Hammer, saw me and said, "I know you like puzzles, so I figure I ought to charge you 15 times the cost of one number since you can make 15 different numbers with 1, 2, 3. But if you can tell me how many numbers you can make with a set of all 10 digits, I'll let you pay the ordinary price for your 123."

Can you figure out his figuring?

189. THE DIE IS CAST

Big Julie has again brought his own dice to Nathan Detroit's crap game. This time they are perfectly normal. However, he has also brought his own table, which you would have to see to believe. It's corrugated with little cubical corner indentations so that the dice won't lie flat. They always come to rest with a corner in one of the indentations and the opposite corner up.

Big Julie says this isn't really a problem — you simply take the sum of the three faces that one can see — OK? And when Big Julie says OK, you say OK.

What values can occur on a single die when playing on Big Julie's table? If we play craps, we have to take the sum of the values on two dice. What values can occur in playing craps on Big Julie's table? On an ordinary table, the probability of the sum of two dice increases linearly to a maximum (at 7) and then decreases again. Does this happen on Big Julie's table?

190. SEX AND THE HONEYBEE

The common honeybee has a most uncommon sex life. The queen first lays unfertilized eggs which develop into drone males. Later she lays fertilized eggs which develop into females, either workers or queens, depending on their diet. Consequently a male has no father, only a mother, while a female has two parents. How many ancestors, broken down by sex, does a queen bee have ten generations ago?

191. FERRYING FIVE

You all know the problem of the farmer with a wolf, a goat and some cabbages wanting to cross a river in a boat that is only big enough to hold the farmer and one of the others. Such problems first occur in a 9th-century manuscript collection entitled *Propositions for Sharpening Youths*, attributed to Alcuin of York. There have been various attempts to extend this problem. A 16th-century version has four animals (who can all row!): dog, wolf, sheep and horse, which we denote as *D*, *W*, *S*, *H*. Each of them dislikes the animals adjacent to him on the list so much that they cannot be left alone together. In 1932, Hubert Phillips ("Caliban") gave a different extension involving a hunter and four bearers. Here is my formulation.

A farmer and his small daughter were traveling with a wolf, a goat and some cabbages. They arrive at a river where there is a boat which only holds two of these. The farmer can control the wolf and the goat, but his daughter is too small to do so, though she can row. Hence the farmer cannot leave the daughter with the wolf, nor the wolf with the goat, nor the goat with the cabbages, unless he is present. How do they get across the river?

192. ASSORTED VOLUMES

On my reference shelf are a number of books of different heights. My wife thinks the room would look better if the books were arranged in order of increasing height. To do this, I can take out one book at a time and re-insert it in the shelf with a bit of sideways pushing. The books are too big for me to remove more than one at a time. How many books do I have to take out and re-insert in this way to get the books in order? Of course you will say that this depends on the order — if they are already in order, I don't have to move any. Quite so, but I want you to figure it out in general. That is, given a particular order, how do you determine how many moves are required? What order(s) requires the most work?

193. DOUBLES MIX-UP

Four married couples, the Adams, the Browns, the Campbells and the Douglases were on holiday at the beach and eventually discovered they were all keen bridge players, so they agreed to play two tables for each of the three remaining nights of the holiday. Variety being the spice of life, they agreed to play mixed pairs, but no one should play with or against his or her spouse. Mr. Adams thought it might even be possible to arrange things so that no one should play with the same person twice nor against the same person twice. Mrs. Brown didn't think this seemed possible at all. But Mr. Campbell pointed out that there were three other spouses to play with and six in the other couples to play against, which were just the right numbers. Mrs. Douglas said that she'd been trying to draw up such an arrangement but hadn't yet succeeded and suggested that the rest of them should get busy and try to make up a playing arrangement. They set to and for a while it looked like they wouldn't get around to playing bridge at all, but suddenly several of them exclaimed: "I've got it!" almost simultaneously. They then compared notes and found they had several solutions, though some were really the same, just with the names permuted. Nonetheless, they found all the possible essentially distinct solutions. Can you?

194. A CHAIN GAME

Readers will know that I spend a lot of time browsing through old puzzle books (and I am grateful for any that you don't want!). I found the following version of a Sam Loyd problem in a 1957 book. I have 9 segments of a chain which I want to have joined into one length. My blacksmith says it will take him 5 minutes to open a link and 10 minutes to weld it closed again. How long will it take him to produce the single length? The solution uses the lengths of the segments, but the author neglected to include the lengths in the problem. So let's try the problem anyway — what is the minimum possible time for joining up 9 segments? And when can you do it in this time? Can you generalize to n segments?

195. IN A TEARING RAGE

The other day I got very frustrated with some calculations that wouldn't come out, so I very neatly folded the piece of paper in half twice and tore it in half and threw it in the rubbish bin. Needless to say, a few minutes later I realized that it was actually all right and I had to fish out the pieces and fit them back together. How many pieces might there have been and how big were they?

[HINT: There are several possibilities — can you find them all?]

196. NOT SO LIKELY

A 1977 magic book gives the following 'Ace — Two — Jack' bet (or swindle). Your opponent (or dupe) is asked to cut the cards into three packs and you bet him that there will be an Ace, a Two or a Jack on the bottom of one of the packs. The author asserts 'you should win about two of every three games'. By now, you should know better than to believe everything you read, so I trust you will work out the odds and see whether 'two out of three' is really likely. Perhaps you can explain what the author had in mind.

197. ELECTION SPECIAL

Jessica and Rachel were putting up a school display about the election. This was to be headed ELECTION SPECIAL and they had stenciled the letters on cardboard and cut them out when Rachel asked whether the letters would fit on the display board. They picked up the letters and laid them out across the top of the board. But since they had used the stencils in alphabetical order, they got ACCEEEIILLNOPST — but there was just room left over for the space between words. They thought this was a pretty funny message and started rearranging the letters looking for other messages. Jessica wondered how long it would take them to try all the messages. Rachel said it depended on how fast they worked. She reckoned that they might do one a second and it shouldn't take too long. Jessica thought it might be pretty long, maybe as much as a year. How long would it take them?

198. DOUBLE JUMPING

You probably know the puzzle where one puts 8 matches or counters in a row and is allowed to move a single counter over two counters and set it down to make a pile of two. The object is to make four such moves and leave four piles of two. With 8 counters the solution is essentially unique — can you find it? How many equivalent solutions are there? The puzzle is sometimes given with 10 or 12 or ... counters, but this is easy — can you see why? However, Jerry Slocum, the American puzzle collector, has an example called Double Five from about 1890 where the 10 counters are arranged in a circle and one wants the final five piles of two to alternate with empty locations. Let us number the original locations of the counters by 1, ..., 10 and say we want the piles of two to wind up in the even locations. Can you do it? Can you do it with 8 counters? Can you arrange to have the final piles of two in consecutive locations?

199. AROUND THE ENVELOPE

It is a favorite childhood puzzle to ask someone to draw a path over all the lines of the 'envelope' pattern below, with each line covered just once.

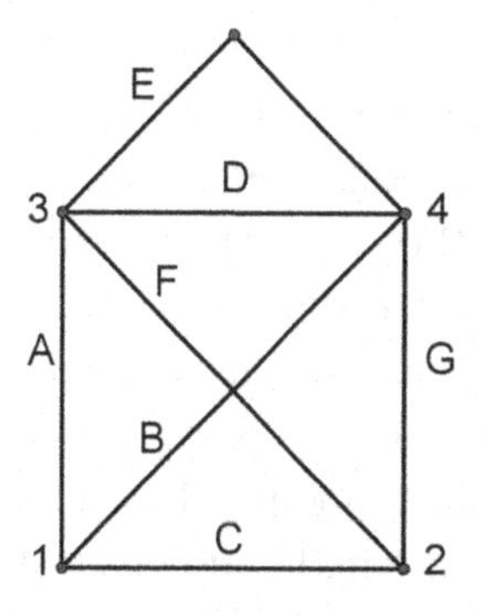

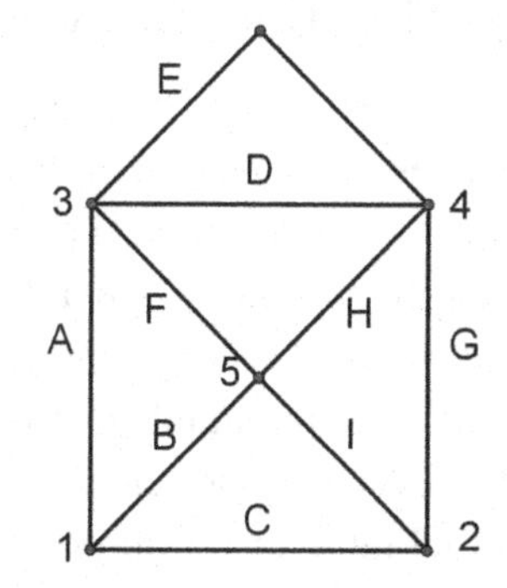

With a little effort, you will find such a path. You may even know or be able to see that the end points of such a path can only be two of the possible corners. Which ones? About five years ago I found a puzzle book that claimed that there are 50 different such paths. I tried doing this by hand and I didn't get that number. How many can you find? For ease of comparison, use the labeling of edges shown above.

As an added complication, I see that there are two ways to view the pattern. My initial view was that there were four points: 1, 2, 3, 4, joined by all six possible connections, with the extra connection E drawn from 3 to 4. In this view, the center point where lines B and F cross is not an actual connection and one cannot shift from line B to line F at this point and this is what the left-hand diagram is intended to convey. But the picture does have this point in it and perhaps one should consider the situation as shown in the right-hand diagram. However, I don't get 50 for either case!

I finally got around to writing a program to find all these paths and my result for the first case was correct, but I had missed some paths in the second case.

200. PLANE COLORING

Mathematicians consider some peculiar questions, both plane and fancy. The famous Four-Color Problem for maps led to considering coloring of other structures. Suppose we try to color the points on the plane so that any two points which are an inch apart have different colors. Show that you need at least four colors. [It is known that you can do it with seven colors, but the minimum number required is not known.]

201. A FURTHER CHAIN GAME

One of the pleasures of creating new problems is that readers sometimes find mistakes or interesting generalizations. Alastair Summers of Stamford, Lincs., looked at the Chain Game problem (Problem 194 above) and asked himself a different, more general, and more interesting, question, which he then solved. Given n segments of chain, having lengths $a_1, a_2, \ldots, a_n$, what is the minimal number of links which must be opened and closed to make this into a single length? In the Chain Game problem, I asked for the minimal number for all possible sets of n lengths and this occurs just for certain sets of lengths. Mr. Summers wants to determine the minimal number for any particular set of lengths. For convenience, assume the a_i are arranged in ascending order, i.e. $a_1 \leq a_2 \leq \cdots \leq a_n$. The answer will depend on the lengths in some way and is a somewhat more abstract problem than usual in this chapter, but I am sure many of you will be able to solve it. Mr. Summers is planning to use it with a secondary school class.

202. THE BLIND ABBESS AND HER NUNS

In 1993, I presented a program on puzzles in Radio 5's "Maths Miscellany" series. In this program, we dramatized an ancient problem, apparently of Arabic origin, generally known as The Blind Abbess and Her Nuns. There is an arrangement of nine rooms in a three by three square. The Abbess occupies the center and there are three young nuns in each of the other eight rooms, making 24 in all. The Abbess, being both blind and dim, checks the attendance each night by counting the number along each side. That is, if A, B, ..., denote the numbers in the rooms as illustrated, then she checks that:

$$9 = A + B + C = C + D + E = E + F + G = G + H + A.$$

$$\begin{matrix} A & B & C \\ H & & D \\ G & F & E \end{matrix}$$

However, the young nuns do a lot of gallivanting and some nights some are away, while other nights they have company. But they always reorganize themselves so the Abbess still finds nine along each side. What numbers of people can be in the eight rooms without the Abbess noticing anything amiss? The most common solutions have all the corner numbers equal and all the side numbers equal — what numbers of people can be accommodated then? Can you generalize to the case where there are a total of S along each side?

A harder problem is to determine all the different arrangements with a constant sum S along each side, or to determine the arrangements which are not equivalent under symmetries of the square. For $S = 9$, the above case, there are 2035 arrangements, of which 365 are really different. These are a bit too large to ask you to find, but can you find the numbers for $S = 2$?

203. YARBOROUGH, YAROO!

In Much Puzzling, the old timers still play a mean hand of whist. Recently Mrs. Baker (the brewer's wife and sister to Mr. Butcher) heard that the Earl of Yarborough would bet £1000 to £1 that a whist hand would have no card higher than a nine. Mrs. Baker told her whist table about this the next time they met. "Cor, I'd take him up on that," said Mr. Brewer, "I get hands like that all the time." Mrs. Butler, the banker's wife, said: "Let's see, there are nine cards lower than a ten, so the probability is 9/13 to the 13th, and . . . ." She fished out her calculator and continued, "that's 8.39 times ten to the minus three. Why that's nearly 1%. He must lose a lot of money!" "And you have to multiply by four for the four hands," said Mr. Brewer. "Just like my husband said, those hands happen all the time," added Mrs. Brewer. "Ah, um," broke in Mr. Butler, the banker, "An ace doesn't count as a card less than ten, so the probability is 8/13 to the 13th, which is only 1.815 times ten to the minus three, which is about one out of 551, so he's not losing so much." "I can't believe he would keep making the bet if it was a loser," said Mrs. Butler. "I get lousy hands much more often than that," blustered Mr. Brewer.

Eventually the debate (or argument) got so persistent that they adjourned their game — an unheard-of event in Much Puzzling — and they all came over to my house. Once I managed to make sense out of all four explanations being given and I managed to get a few words in edgewise, I told them that the relevant Earl wasn't taking any more such bets as he was dead, indeed had been dead over a century, that such a hand is generally known as a Yarborough, that the bet is for just one hand, not for the four hands of a deal and that their calculations weren't very good. So what are the right odds? Did Yarborough make or lose money on his wager?

204. AN UNLIKELY START TO A CHESS GAME

Jessica and her friend Rachel have been playing a lot of chess with our old chess set. When they were setting up the board today, Jessica noted that the pieces were so marked up that they were all different, if one looked closely. Rachel said that meant that you could tell which pawn was which, and she was going to set out her men in a different way each time. Jessica asked if she had any idea how long she could play that way. Rachel said she didn't know but she bet it was at least a year. Jessica said she thought it would be longer, several years at least. How many different ways can one player set out her pieces? How many ways can two players set out their pieces? How long will it take for them to play through all such games?

205. WIMBLEDON WORRIES

When Wimbledon comes around, my family gets involved in arguments about possible scores. As an easy example, how many different scores can occur in a set? Actually the number is infinite unless the tie-break rule is used, so we suppose that the tie-break rule is used in all sets.

What really annoys us is that in many matches, the losing player can win more games than the winner. How many more games, at most, can a loser win?

We also notice that both players can win the same number of games. There are many ways this can happen, but what are the minimal number of games and the maximal number of games that can occur? How many sets can occur in such games?

You may as well do both the ladies' and the men's cases. Assume that no extraordinary events occur, such as defaults.

206. DOUBLE JUMPING — 1

Most readers are probably familiar with the standard double jumping problem — see Problem 198 above. One sets out 8 (or 10 or 12 or ...) counters in a row and one can move a single counter over two other counters onto a single counter to produce a pile of two counters. The object is to produce 4 (or 5 or 6 or ...) piles of two counters. It is not hard to solve this for any even number $\geq$ 8 and the number of solutions rapidly gets large, being 16, 48, 944, ... Reading an early 20th-century book, I found two variations that I have not seen elsewhere. The first imagines that the counters are numbered 1, 2, 3, ..., e.g. by using cards, and asks for the solution with the minimum total of exposed counters, i.e. the minimum total of the top counters, which are just the moved counters. Can you find this minimum value for the first few cases: 8, 10, 12?

207. DOUBLE JUMPING — 2

Continuing from Double Jumping — 1, the early 20th-century book provided a second novel variation. It moves the counters over two <u>piles</u>, rather than over two counters. Show that this permits solutions for any even number ≥ 6. Again, the number of solutions gets large — for 6, 8, 10, the number of solutions is 8, 60, 456. Can you find solutions with minimal total of exposed counters for 6, 8, 10? [For 10 counters, my source said 20 was minimal — can you do better?]

208. SINGLE JUMPING

Following on from Double Jumping — 2, I observed that in this case, one could also consider jumping over one pile. This leads to an acceptable problem, whereas jumping over one counter is clearly an unsolvable problem. Show that jumping over one pile is solvable for any even number ≥ 4. For 4, 6, 8, 10, there are 4, 16, 144, 1408 solutions. Find solutions with minimum total of exposed counters for 4, 6, 8, 10.

209. STAR BRIGHT

In a Christmas puzzle book of a few years ago, I found the following pattern with the instruction "How many triangles can you find . . . ?" The solution is "At least fifty". This seems like a pretty feeble answer to me, so perhaps you can do better and really find all the triangles.

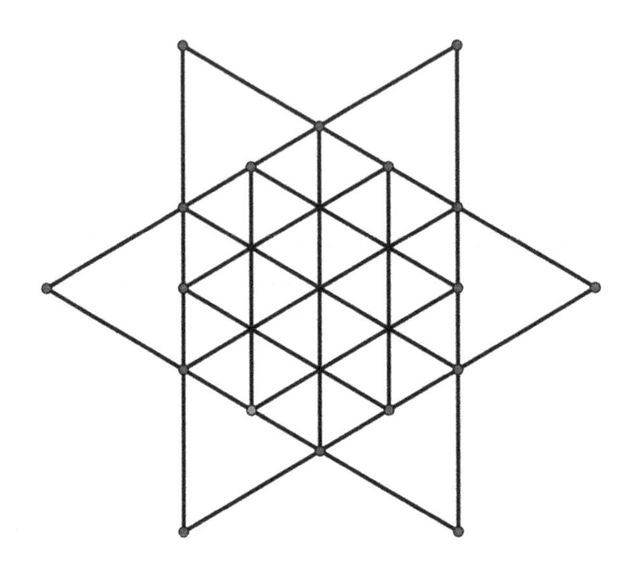

210. ANOTHER TRIANGLE NUMBER

I have just found another example of counting triangles in a figure. This comes from a 1939 puzzle book and the only older examples that I have found are from 1907, 1908 and 1928. Draw an equilateral triangle ABC and let a, b, c denote the midpoints opposite to A, B, C respectively. Take a point d between a and B. Draw the lines AB, AC, Aa, BC, ab, ac, bc, bd, cd. How many triangles are there in the figure? The answer given was 24, but I can find more than that and I trust you can too.

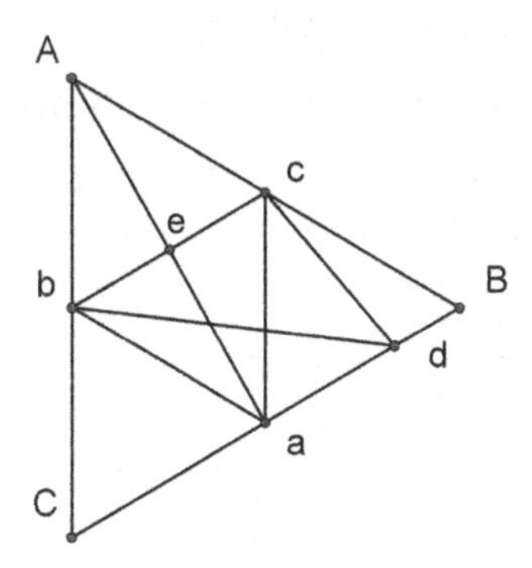

211. PICTURE CARD PROBABILITY

Reading through a popular magic book, I find the following trick. Place the twelve picture or court cards together. Ask someone to put them anywhere in the deck he likes and then cut the cards twelve times. You then show that the twelve cards are still together — but, adds the author, "there's one chance in five hundred that this will go wrong". Having seen some problems, you undoubtedly now know enough to take such assertions with a large grain of salt. So, what is the real probability? How does it depend on the number of cuts?

212. PAINTED CUBES

A wood block of integral dimensions $A \times B \times C$ is painted on all sides. It is then cut into unit cubes in the natural way. Is it possible for exactly half of these unit cubes to be completely unpainted? How many such blocks are there? What are the largest and smallest examples?

213. A PLACING PROBLEM

Consider a standard 3 by 3 board. I have nine markers, three each of three different colors. I want to place them on the board so that each marker is adjacent, either horizontally or vertically, to markers of both other colors. I recently found this in a book from the 1930s which claimed there was one solution. I think you can immediately see that one solution implies several others, so how many solutions are there and are there any that are really different?

214. HOW MANY CHESS PLAYERS?

Chess tournaments are sometimes divided into two classes and each player in a class plays one game against every other player in his class. In a recent tournament with two classes, 100 games were played. How many players were there?

[From Hubert Phillips; *News Chronicle "Quiz" No. 3: Natural History*; News Chronicle, London, 1946, pp. 11 & 33.]

Chapter 15

Some Verbal Puzzles

Although this book is about mathematical puzzles, there are problems which are based on words, but use the kind of thinking that mathematicians employ.

215. MATCHSTICK WORDS

The other day Jessica was fiddling with a box of matches. She used the matches to spell the name of her friend Anna and then tried to spell other names, but most names had curved letters and really didn't look right with matches. Then she tried to form ordinary words, trying to find a really long word. How long a word can you find?

For definiteness, we must agree that only the following letters are usable:

A, E, F, H, I, K, L, M, N, T, V, W, X, Y, Z.

216. A JOLLY GATHERING

At a recent gathering of the Much Puzzling Metagrobological Society, Mrs. Baker brought out an item she found in a recent American puzzle book. "See the following letters: ERGRO." Mrs. Brewer butted in: "I know that word. That's what that French card-player said; 'Coito, ergro some', isn't it?" After we had all recovered from this and explained about Descartes, Mrs. Baker continued: "The object is to place three letters in front and the same three letters in back, like ABCERGROABC, in order to make a common word. Mr. Butcher looked up from his crosswords and mumbled some — cross words, that is — which sounded like 'troglodytes' to me. I would swear he said 'troglodytes' — and Mrs. Brewer must have heard the same thing as I heard her mumble the word and say she didn't see how that could fit. "Eureka, I've got it", yelled Mr. Butcher. "What?", we all demanded. "The question is pretty well known," he responded, "so I've been looking for other versions of the problem. I've found a version where you put three letters in front and the same three letters in back, but not in the same order. So it might look like ABCERGROCBA — or then again it might not." The collected metagrobologists racked their brains. Can you help them out? And what was the easy solution?

217. AN INTERMEDIATE PROBLEM

We know that grey is between black and white and twilight is between day and night, but what's between up and down?

We also know that two's company and three's a crowd, so what are four and five?

218. CALCULATED WORDS

When you hold your calculator upside down, most of the digits can be read as letters. The digits 1, 2, 3, 4, 5, 7, 8, 9, 0 can be read as I, Z, E, H (in lower case as h), S, L, B, G, O.

(This is clearer on some calculators than others and some might say the upside-down 6 is closer to a g than upside-down 9 is to G.) A common amusement has been to produce words in this way. What is the longest word that you can produce?

Last time I asked a question like this, I got lots of answers much longer than those I had found and I suspect you will again rise to this challenge. I have found 42 7-letter words; 26 8-letter words; 13 9-letter words; 5 10-letter words; 1 11-letter word; 3 12-letter words and 1 13-letter word. Some of the words look better with a hyphen, but of course that is an upside down — sign. But then one ask if one can do better allowing an *x* which is an upside-down times sign — though some calculators show a *.

219. UNNUMBERED LETTERS

Which letters never occur in the name of any positive whole number (at least as far as is known)? Make a list of the letters and the first numbers in which they occur. You will have to go further than you might expect.

220. ALOHA

The Hawaiian alphabet has only 12 letters: A, E, H, I, K, L, M, N, O, P, U, W. As a result, the Hawaiians use long words — e.g. they had kings named Kamehameha and a queen named Liliuokalani — and some repetition — e.g. mahimahi is the fish called a dolphin, not to be confused with the marine mammal of the same name. So what is the longest English word that can be formed using the Hawaiian alphabet? Since it is possible that this is a Hawaiian word that has recently become part of English in Hawaii or California, it seems best to rule out words of Hawaiian origin.

221. A VERY SNEAKY SEQUENCE

What is the next number in the following sequence?

15145, 202315, 2081855, 6152118, 69225, ...

Solutions

Chapter 1

1. SHARE AND SHARE ALIKE

Since they shared the sandwiches, they each ate three sandwiches, so each sandwich is worth \$1. So Jessica gave one sandwich to Samantha while Pud gave two, so Jessica should get \$1 and Pud should get \$2.

2. LEO'S LILLIAN LIMERICK

The first problem is to find $S = 1 + 2 + \cdots + 1{,}000{,}000{,}000$. You may already know this as the sum of an arithmetic progression. If not, the easiest way to find it is to consider it backwards:

$$S = 1{,}000{,}000{,}000 + \cdots + 2 + 1.$$

If we now add the corresponding terms in the two sums, we find that $2S$ is a sum of a billion terms all equal to 1,000,000,001. So

$$S = 500{,}000{,}000 \times 1{,}000{,}000{,}001 = 500{,}000{,}000{,}500{,}000{,}000.$$

Using powers of ten notation, we can write this a bit more simply as $5 \times (10^{17} + 10^{8})$.

The second problem, though it stumped Lillian, is actually much easier. Consider the numbers from 0 to 999,999,999. Write all of them with nine digits, so the numbers start 000,000,000; 000,000,001; 000,000,002 and end at 999,999,999. In each digit position, the ten digits $0, 1, 2, \ldots, 9$ occur equally often, hence they appear 100,000,000 times in each position. Thus the sum of the digits in any position is

$$\begin{aligned} 100{,}000{,}000\ (0 + 1 + 2 + \cdots + 9) &= 100{,}000{,}000 \times 45 \\ &= 4{,}500{,}000{,}000. \end{aligned}$$

There are nine digit positions and we also have to include the sum of the digits of 1,000,000,000, so Lillian's solution is $9 \times 4{,}500{,}000{,}000 + 1 = 40{,}500{,}000{,}001$. So Lillian really owes you forty billion.

[If you prefer the older English usage where a billion is a million million ($1{,}000{,}000{,}000{,}000 = 10^{12}$) instead of a thousand million

$$(1{,}000{,}000{,}000 = 10^9),$$

then the answers are

$$\begin{aligned} &500,000,000,000,500,000,000,000 \\ &\quad = 5 \times (10^{23} + 10^{11}) \quad \text{and} \quad 54,000,000,000,001.] \end{aligned}$$

3. SOME SQUARES

Let a, b, c, d denote the numbers of items bought. Then we want $a^2 + b^2 = c^2 + d^2$, with distinct numbers a, b, c, d. A little searching finds the first solution: $1^2 + 8^2 = 4^2 + 7^2 = 65$. So each couple spent \$65, making \$130 in total.

[There is a smaller case: $1^2 + 7^2 = 5^2 + 5^2 = 50$, which is ruled out by requiring all numbers to be different, although it genuinely represents 50 as a sum of two squares in two different ways. The problem has also ruled out exceptional cases such as $0^2 + 5^2 = 3^2 + 4^2 = 25$.

Exercises for the ambitious — find the smallest number representable as a sum of two cubes (or two fourth powers, etc.) in two different ways. I don't know if this has been done for fifth powers.]

4. SOME PRODUCT

Suppose we have just two numbers, a and b. We can assume $a \leq b$. We want $a + b = ab$. Solving this for b gives $b = a/(a-1)$. Trying $a = 1, 2, \ldots$, we see that b quickly decreases below a, and there is only one solution: 2, 2.

Now try with three numbers, $a \leq b \leq c$. Then $a + b + c = abc$ gives us $c = (a + b)/(ab - 1)$. We systematically try a, $b = 1, 1; 1, 2; 1, 3; \ldots$ until $c < b$. Then we try a, $b = 2, 2; 2, 3; \ldots$ until $c < b$, etc. We get only one solution: a, b, $c = 1, 2, 3$.

Continue with four numbers, $a \leq b \leq c \leq d$. Then $a + b + c + d = abcd$ gives us $d = (a + b + c)/(abc - 1)$. Systematic trial again finds just one solution: $a, b, c, d = 1, 1, 2, 4$.

Continue with five numbers. We get $e = (a + b + c + d)/(abcd - 1)$. Trial now reveals three solutions:

$$a, b, c, d, e = 1, 1, 1, 2, 5; \quad 1, 1, 1, 3, 3; \quad 1, 1, 2, 2, 2.$$

So the original list must have contained at least 5 numbers and there are three possible lists with 5 numbers.

[I continued my search and found that there is just one list with sum equal to product with 6 numbers: 1, 1, 1, 1, 2, 6, but it appeared that there always were two or more lists with sum equal product with 7 or more numbers. I put this on my computer and was surprised to find that there are unique lists of n numbers with sum equal product for $n = 24$ or 114 or 174 or 444. There are no more such n up to 7550, but I could not prove that there are no more. Perhaps some reader can.]

5. SUM TROUBLE

Hannah is right — there is no way to insert just plus signs into the sequence to add up to 100. One uses the ancient technique of 'casting out nines' or the more modern equivalent of congruence (mod 9).

[For those who don't know or don't remember, the idea is that any number is congruent to the sum of its digits (mod 9) and that the arithmetic operations $+$, $-$, $\times$ are preserved by this congruence. For example, 21 is congruent to 3 (mod 9) and 32 is congruent to 5 (mod 9), so $21 + 32 = 53$ is congruent to $3 + 5 = 8$ (mod 9). If the sum of the digits is greater than 9, we can repeat the process, e.g. 58 is congruent to $5 + 8 = 13$ (mod 9), which is congruent to $1 + 3 = 4$ (mod 9). We also have that 9 is congruent to 0 (mod 9). Now we also have $21 - 32 = -11$ is congruent to $3 - 5 = -2$ (mod 9), which is congruent to 7 (mod 9) and $21 \times 32 = 672$ is congruent to $3 \times 5 = 15$ (mod 9), which is congruent to $1 + 5 = 6$ (mod 9). Another way to form the sum of the digits is simply to 'cast out a nine' each time the sum exceeds 9. For example, we consider 672 and say $6 + 7$ is 13, cast out 9 leaves 4, $4 + 2$ is 6.]

How does this apply to Jessica's problem? Quite easily — no matter how we insert $+$ signs, the terms will contain all the digits $1, 2, \ldots, 9$ and the sum of the terms will hence be congruent to the sum of these digits (mod 9). But $1+2+\cdots+9 = 45$, which is congruent to 9 or 0 (mod 9), while 100 is congruent to 1 (mod 9). So no sum with digits $1, 2, \ldots, 9$ can add up to 100, not even if we reorder them. Observe that the solution given had a -4 instead of a $+4$, which had the effect of subtracting 8 and so the sum of the numbers and the sum of the digits comes out $9-8 = 1 \pmod 9$ as required.

6. RUSH AND SEDGE

The Chinese approach was to look at the sizes at the end of each day. This gives the following sizes.

Day	1	2	3	4
Rush	3	$4\frac{1}{2}$	$5\frac{1}{4}$	$5\frac{5}{8}$
Sedge	1	3	7	15

Clearly the sedge equals the rush sometime during the third day. If we assume that the plants grow at constant rate throughout the third day, then the sizes of the two plants at time t into the third day are: $9/2 + 3t/4$ and $3+4t$. Here t is measured in units of days. Setting these two quantities equal and then collecting like terms, we obtain $13t/4 = 3/2$, whence $t = 6/13$ and so the total time is $2\ 6/13 = 2.4615\ldots$.

However, it seems clear that the plants are growing at rates which vary continuously throughout the day. One can apply calculus to this, but one can also proceed a bit more intuitively. Consider first the sedge. After n days, it will have grown $S(n) = 1 + 2 + 4 + \cdots + 2^{n-1}$. This sum is a geometric progression and can be seen in many ways to be equal to $2^n - 1$. In school algebra, one learns that 2^t is defined for any real value of t as a natural generalization of the values for integer t. Thus it is fairly natural to assert that the sedge has size $S(t) = 2^t - 1$ after t days, and the use of calculus gives the same result. Now in n days, the rush has grown

$$\begin{aligned} R(n) &= 3 + 3/2 + 3/4 + 3/8 + \cdots + 3/2^{n-1} \\ &= 3[1 + 1/2 + 1/4 + \cdots 1/2^{n-1}]. \end{aligned}$$

Again this is a geometric progression, and it is easy to see directly that the sum in brackets is just $2 - 1/2^{n-1}$. Hence

$$R(n) = 6 - 3/2^{n-1}$$

and we can take this as defined for all real t just as with the sedge.

Thus we want the time t such that $R(t) = S(t)$, i.e. $6-3/2^{t-1} = 2^t - 1$. If we set $x = 2^t$, we obtain $6 - 6/x = x - 1$, which is a quadratic equation which can be solved by the quadratic formula. But it is easier to see that the left-hand side is $6(x-1)/x$, so that canceling the $x-1$ leaves $6/x=1$ or $x=6$. Using logarithms, we can write the solution as $t=\log_2 6=(\log 6)/(\log 2)$, where the first log is to the base 2, while the later logs are to any base. Numerically, we get $t=2.58496\ldots$.

[In canceling the factor $x-1$, we are assuming that this factor is not zero, i.e. that $x\neq 1$. However, this is the same as saying that $t\neq 0$, which points out that there is a second, trivial, solution at $t=0$, when both plants have not grown at all!]

7. SHE'S A SQUARE

The only square year in the near future is $45^2=2025$. Hence Katie was born in 1980 and she will be 36 years old in 2016.

8. LEMONADE AND WATER

Zero! The amount of lemonade gained by Jessica is precisely the amount of lemonade lost by Rachel. But Rachel has just as much liquid afterwards as before, so the lost lemonade must be replaced by an equal amount of water.

9. AN OLD MISTAKE

A standard form for an arithmetic progression of five terms is: a, $a+d$, $a+2d$, $a+3d$, $a+4d$. The sum is $5a+10d$, which is 5 times the middle term. Since the sum is 40, we know the middle term $a+2d$, is 8. It is now easier to write the progression symmetrically as: $8-2d$, $8-d$, 8, $8+d$, $8+2d$, and the product is $8(64-d^2)(64-4d^2)$. Setting this equal to 12320, we can cancel 8 and 4 to leave $(64-d^2)(16-d^2)=385$. Letting $x=d^2$, this is a quadratic equation for x, namely $x^2-80x+639=0$. The roots are $x=9$ and $x=71$. Both answers work, but only $d=3$ keeps the terms all positive, giving 2, 5, 8, 11, 14. So the answer should have been 2 and 3 is the answer if the common difference is asked for.

[Incidentally, $d=-3$ and $d=-\sqrt{71}$ are also legal solutions. They just give the same five terms as for the positive ds, but in reverse order.]

10. HORSE TRADING

Suppose the trader bought h horses at price p per horse. Then $hp = (h-1)(p+20)$, which yields:

(1) $0 = 20h - p - 20$.

If Jessica is right, then $hp = (h-2)(p+40)$, which yields.

(2) $0 = 40h - 2p - 80$, which is inconsistent with equation (1).

Hence Jessica is wrong and Hannah must be right. Hannah's statement gives us $hp = (h-2)(p+45)$, which yields.

(3) $0 = 45h - 2p - 90$. Subtracting twice (1) from (3) leaves us $0 = 5h - 50$, whence $h = 10$ and (1) then gives us $p = 180$.

11. A WEIGHTY PROBLEM

With weights $3, 4, 6, 27$, one can obtain any weight up through 40, EXCEPT $8, 11, 12, 15, 16, 19, 35, 38, 39$. With weights $3, 4, 9, 27$, one can obtain any value up through 43 EXCEPT $38, 41, 42$, so our whey-weigher can weigh 8 more weights up through 40. The only set of weights that will work is $1, 3, 9, 27$.

[Suppose we have four weights A, B, C, D, in increasing order. The values that we can weigh out on a balance are represented by $aA + bB + cC + dD$, where a, b, c, d can take on the values $+1$ (by using the weight in the ordinary way (or weigh?)), 0 (by not using the weight) and -1 (by putting the weight in the pan with the whey). Thus there are $3^4 = 81$ possible ways to weigh whey. Of these, one trivial way is to use no weights, which gives value 0, leaving 80 non-trivial weighings. By reversing the weights in the pans, we see that each weighing has a corresponding negative, and there are thus at most 40 positive weighings. [(The 'at most' arises since some of the non-trivial weighings might have value 0, e.g. if some weight occurs twice in our set.) Similarly, with a set of n weights, the maximum number of positive weighings is $(3^n - 1)/2$.]

[Thus, with four weights, we can at most weigh out $1, 2, \ldots, 40$ and we must have $A + B + C + D = 40$. Now consider how we could weigh out 39. This can only be obtained as $B + C + D$, whence we deduce that $A = 1$ and $B + C + D = 39$. By moving the 1 to the other side, we can weigh out 38, hence we do not need a weight of 2, i.e. $B > 2$. Now consider weighing

37. The next smallest value that we can obtain is $A + C + D = 40 - B$, so we must have $B = 3$ and $C + D = 36$. The various ways of placing or removing 1 and 3 from 36 give us the weights 32, 33, 34, ..., 40. From this we deduce that $A + B + D = 40 - C = 31$, so $C = 9$ and $D = 27$.]

12. TRIANGULATION

There is no easy way to solve this. One must study the 'Triangular Numbers'. The n-th triangular number is $T(n) = 1 + 2 + 3 + \cdots + n$, which is well known to be $T(n) = n(n + 1)/2$. The problem says that the number of marbles is a triangular number which is divisible by precisely eight triangle numbers. If $T(m)$ divides $T(n)$, then $m(m+1)$ divides $n(n+1)$ and so both m and $m + 1$ are divisors of $n(n + 1)$. This gives a systematic way of finding all the triangular numbers which divide a given triangular number. For example, $2 \times T(8) = 72$ which has the divisors 1, 2, 3, 4, 6, 8, 9, 12, 18, 36, 72, so $T(8)$ is divisible by $T(1) = 1$, $T(2) = 3$, $T(3) = 6$ and $T(8) = 36$. Continuing in this way, we find that $T(20) = 210$ is divisible by $T(m)$ for $m = 1, 2, 3, 4, 5, 6, 14, 20$. The next cases which have exactly eight 'triangular divisors' are $T(90) = 4095$ and $T(95) = 4560$. $T(44) = 990$ and $T(80) = 3240$ have nine triangular divisors and $T(35) = 630$, $T(84) = 3570$ and $T(99) = 4950$ have ten triangular divisors.

Perhaps some reader can find a formula for the number of triangular divisors. I have not found one.

13. MATCHSTICKS, OR FIDDLESTICKS!

Here are my solutions. You probably found some others.

0. II −II
1. II/II or −I/−I
2. II × I
3. IV − I or II + I or − − III
4. IV/I
5. V × I
6. V + I
7. IIIX or VII^{I} or $\sqrt{}$ IL, i.e. the square root of 49
8. VIII
9. IX/I or III^{II}, i.e. three squared or 11 − II, i.e. eleven minus two

10. XI − I or X × I
11. X + I or XI/I or 11 × I, i.e. eleven times one
12. 11 + I, i.e. eleven plus one
13. XIII
14. XIV
15. XV^{I}
16. XVI
17. 17/1, where the seven is made from two matchsticks — if you don't like this, you can make 17 where the one and the stem of the 7 are made from two matchsticks.

[This is adapted from the *Morley Adams Puzzle Book* of 1939 which allowed 4 − I for three and 10 for ten by viewing 4 and 0 as made from four matchsticks.]

14. CALCULATED CONFUSION

Let our three numbers be A, B, C. The first method computes $A + B \times C$ as $(A+B)C = AC + BC$. The second method computes $A + BC$. The two give the same result if and only if $AC = A$, which is if and only if $A = 0$ or $C = 1$. The first is larger than the second if and only if $AC > A$, i.e. $AC - A > 0$, i.e. $A(C-1) > 0$, which holds if and only if ($A > 0$ and $C > 1$) or ($A < 0$ and $C < 1$). The second result is larger if and only if ($A > 0$ and $C < 1$) or ($A < 0$ and $C > 1$). Jessica's problem was that she was only trying positive integers and thought 1 was too special to be very interesting.

15. JELLY BEAN DIVISION

Let N be the number of beans. Jessica's result says that $N - 3$ is divisible by 7. Hannah's result says that $N - 6$ is divisible by 8 and Rachel's result says that $N - 9$ is divisible by 9, i.e. N is divisible by 9. The Chinese Remainder Theorem (actually due to the Chinese! — its first known appearance is in Sun Zi (Sun Tzu); *Sun Zi Suan Ching* (*Master Sun's Arithmetical Manual*) of the 4th century) gives a general process for solving such problems, but for small values, we can proceed by systematic trial and error. The numbers satisfying Jessica's result are: 3, 10, 17, 24, 31, 38, ... The remainders of these upon division by 8 are: 3, 2, 1, 0, 7, 6, . . . , so that 38 satisfies both Jessica's and Hannah's results. Then the numbers satisfying both Jessica's and Hannah's results are: 38, 94, 150, ... The remainders of these on division by 9 are: 2, 4, 6, ... , so we see we have to go to the 8th term, namely 486, to get one divisible by 9. The general solution is obtained by adding any multiple of 504 (the least common multiple of 7, 8, 9) to 486. Thus the next solution is 990 which is too large.

16. SONS AND DAUGHTERS

Using a calculator, one sees that 49,200 10/13 does not divide exactly into 1,920,000, but gives a result of 39.02.... Since 13 is a factor of 39, it seems clear that the answer is supposed to be the 39th part of 1,920,000 which turns out to be 49,230 10/13, so there has been a simple misprint in the answer.

Each daughter receives twice what the mother does and each son receives six times what the mother does. If there are S sons and D daughters, this means that $6S+2D+1 = 39$ or $3S+D = 19$. This has seven solutions: $S, D = 0, 19;\ 1, 16;\ \ldots;\ 6, 1$. Any of these solutions might be the missing data for the problem.

17. THREE BRICKLAYERS

Let A, B, C denote the amount of the wall that Al, Bill, Charlie can build in one day. Then the data tells us that

$$A + B = 1/12, \quad A + C = 1/15, \quad B + C = 1/20.$$

There are various ways to solve such a system of three equations in three unknowns, but the symmetry of the problem should inspire us to add the equations to get

$$2(A + B + C) = 1/12 + 1/15 + 1/20 = 12/60 = 1/5.$$

Hence $A + B + C = 1/10$ and all three can build the wall in 10 days. Subtracting the original equations from this last equation gives

$$C = 1/10 - 1/12 = 1/60; \quad B = 1/10 - 1/15 = 1/30;$$
$$A = 1/10 - 1/20 = 1/20,$$

so that Al, Bill, Charlie can build the wall by themselves in 20, 30, 60 days.

To make all the numbers come out as integers, we have to have all of A, B, C, $A+B$, $A+C$, $B+C$, $A+B+C$ be fractions with unit numerators. To combine them easily, imagine that all these fractions have been given a common denominator d, so we can consider $A = a/d$, $B = b/d$, etc., and we want $a, b, c, a+b, a+c, b+c, a+b+c$ to all divide d. We can achieve this easily by taking any three integers a, b, c, and letting d be the least common multiple of $a, b, c, a+b, a+c, b+c, a+b+c$. Taking $a, b, c = 3, 2, 1$, we find $d = 60$ and the given problem is the simplest example with distinct rates A, B, C.

18. AN ODD AGE PROBLEM

In y years from now, Jessica will be $y + 16$ years old and Helen will be $y+8$ years old. The ratio of ages is then $R = (y+16)/(y+8) = 1+8/(y+8)$, which is equal to one only when y is infinite, i.e. in the limit as y gets infinitely large. This explains Jessica's remark about my comment being just the limit.

Examining R, we see that it is bigger than one when $y > -8$, i.e. ever since Helen was born. But if we go ahead and look further back, then R is negative for $-16 < y < -8$ and then R is again positive for $y < -16$. Setting $R = 1/2$ gives $y + 16 = y/2 + 4$, whence $y = -24$. So 24 years ago, Jessica was -8 years old and Helen was -16 years old! This shows the positive power of negative thinking!

19. SOME SQUARE SUMS

Clearly we must have 0 in each set to give a sum of 0. No other number can occur in both sets. The number 1 must occur in one of the sets and we will put it in the first set which will be written at the top of the table. As we try to fill in the table, we see that the first number which does not yet occur as a sum must be the next number in one of the sets and this allows us to rapidly work through all possibilities to find the following. $\{0, 1, 2, 3\}$ & $\{0, 4, 8, 12\}$. $\{0, 1, 4, 5\}$ & $\{0, 2, 8, 10\}$. $\{0, 1, 8, 9\}$ & $\{0, 2, 4, 6\}$.

However, there are some other cases. I didn't specify that the two sets had to be the same size. This permits a trivial case of $\{0\}$ & $\{0, 1, \ldots, 15\}$, but there are also cases where one set has two elements and the other has eight. These are easily found to have their two-element sets being: $\{0, 1\}$; $\{0, 2\}$; $\{0, 4\}$; $\{0, 8\}$.

20. A PRETTY PIZZA PROBLEM

Let x, y, z denote the number of employees of the three types and let n be the length of the shift in hours. Then we have $x + y + z = 9$ and

$$[500x + 375y + 135z]n = 33360.$$

We can divide the second equation through by 5 to get

$$[100x + 75y + 27z]n = 6672.$$

Factoring 6672 gives us $6672 = 16 \times 3 \times 139$, so the small factors of 6672 are: 1, 2, 3, 4, 6, 8, 12, 16, 24. (Since this is a daily shift, we must have $n \leq 24$.) Further, the maximum wages occur when $x = 9$, $y = z = 0$, so we must have $900n \geq 6672$, which gives us $n \geq 7+$, so n can only be 8, 12, 16 or 24, giving $6672/n = 834$, 556, 417 or 278. Rewriting our second equation as $100x + 75y + 27z = 6672/n$, multiplying the first equation by 27 to get $27x + 27y + 27z = 243$, and subtracting, we get $73x + 48y = 6672/n - 243$, and the right-hand side takes on the values 591, 313, 174, 35. It is easy to test all possible non-negative values of x in each case and find there is just one solution: $n = 12$, $x = 1$, $y = 3$, $z = 5$.

21. GEE WHIZZ!

The answer is indeed a misprint for 10, as I originally conjectured. But the relationship is one degree more involved than I imagined. Put the 10 into the array.

$$\begin{array}{rrr} 7 & 3 & 11 \\ 9 & 5 & 22 \\ 10 & 2 & 27 \end{array}$$

Then one can see that adding the first two rows and subtracting the third leaves the constant value 6.

I.e. $7 + 9 - 10 = 6$; $3 + 5 - 2 = 6$; $11 + 22 - 27 = 6$.

I only considered that the third row might be a combination of the first two rows and didn't think to allow for a constant row.

22. SUMS OF THREE FACTORS

The factors of a number, n, come in pairs: a and n/a and either of the factors can be viewed as a. For example, $6 = 2 \cdot 3$ with $3 = 6/2$ or $6 = 3 \cdot 2$ with $2 = 6/3$. Since we want to add up our factors and compare them with n, it is easier to consider our factors as having the form n/a. For example, when we write $6 = 6/6 + 6/3 + 6/2$, we can cancel the 6 and we get $1 = 1/6 + 1/3 + 1/2$. From any representation of 1 as a sum of the reciprocals of three distinct integers, we have a representation of n as a sum of three factors, provided n can be divided by each denominator. In our example, this is if and only if n is a multiple of 6. For example, taking $n = 12$, we have $12 = 2 + 4 + 6$.

So now we want to find representations of 1 as a sum of three distinct reciprocals and it is easiest to consider them in decreasing order. We can't use 1/1 as then there is nothing left for the other two terms. If we start with 1/2, the next available value is 1/3 and this works with a third value of 1/6. If we try $1/2 + 1/4$ the third term must be 1/4. If we start with $1/2 + 1/5$, this leaves something larger than 1/5 and hence our fractions are not in decreasing order and this applies for any further example with $1/2 + 1/b$ where $b > 4$. If we start with 1/3, the largest sum we can get is $1/3 + 1/4 + 1/5 = 47/60$ which is less than 1. So the only representation of 1 as a sum of three distinct reciprocals is $1/2 + 1/3 + 1/6$ and a number is a sum of three distinct factors if and only if it is a positive multiple of 6.

It is easily seen that no number is a sum of two distinct factors as this would lead to 1 being a sum of two reciprocals of integers greater than one and the largest such sum is $1/2 + 1/3 = 5/6$.

For four factors, the situation gets a little more complex, but the analysis is identical to that for the famous problem of the 17 camels.

I will just list the denominators of the solutions:

2, 3, 7, 42; 2, 3, 8, 24; 2, 3, 9, 18; 2, 3, 10, 15; 2, 4, 5, 20; 2, 4, 6, 12.

These can be applied to any integer n that is divisible by all of the denominators, i.e. to any multiple of the least common multiple (LCM) of the denominators. The LCMs are: 42, 24, 18, 30, 20, 12. Hence a number is the sum of four distinct divisors if and only if it is a multiple of 12, 18, 20, 30 or 42.

23. SQUARE CONNECTIONS

To begin, the number 1 can be connected to 3, 8, 15, 24, ... , so N must be at least 3. Now 2 can be connected to 7, 14, 23, ... (connecting 2 to 2 doesn't do any good), so N must be at least 7. However, looking at the possible connections among $1, 2, \ldots, 7$, we see that most numbers can only be connected to one other number. If we want to string all the numbers together then at most two numbers can have just one possible connection and they must occur at the ends of our string. So we need to increase N. As we do so, we find that the extra values have only one possible connection until we get to 13 when we start providing extra connections. At $N = 15$, there are just two numbers with just single connections: 8 and 9. So 8 must be connected to 1, but then there is a choice of connecting to 3 or 15. One can try both cases, but it is easier to look at 9 and see that we must have the sequence: 9, 7, 2, 14, 11, 5, 4, 12, 13, 3 with a choice of 1 or 6 as the next term. But 1 has to be followed by 8, so we have to take 6 and the sequence is easily completed to: 9, 7, 2, 14, 11, 5, 4, 12, 13, 3, 6, 10, 15, 1, 8. This is unique up to reversing the string.

This sequence was discovered by Bernardo Recamán, of Bogotá, Colombia, in 1990. I'm grateful to him for sending it to me. He and his colleagues have found that the smallest N for which there is a cycle of this type is 32, and the solution is essentially unique. They also found that there is such a cycle for all larger N up to 1300, and they conjecture that there is such a cycle for all $N \geq 32$. The number of cycles increases rapidly. It seems likely that similar results hold if 'square' is replaced by 'cube' or 'k-th power', but one has to go up to 473 to get a cycle in the cube case.

Chapter 2

24. SHIFTY MULTIPLES

Let the number look like $\ldots cba9$. When I multiply this by 9, the result has last digit (i.e. its units digit) equal to 1, since $9 \times 9 = 81$. But this is to be the same as $9 \ldots cba$, so we deduce that $a = 1$. So our number now looks like $\ldots cb19$. When multiplied by 9, the result ends in 71, so $b = 7$.

Generalizing this, we can compute the digits $a, b, c, \ldots$ as now illustrated for b. Compute $9 \times a$ and add any carry from the calculation of a. This gives a result whose units digit is b. If the result is greater than 9, it gives a carry to the next position. This gives us the following digits: 17422... We can stop once we get a 9 occurring with no carry to it as this is then the same situation as we started in. I will leave it to the grinds among you to verify that the number was:

$$10,112,359,550,561,797,752,808,988,764,044,943,820,224,719.$$

(Strictly, one can repeat this number any number of times to get further answers, but we'll settle for the first answer. Mr. Grind wouldn't give an easy problem where repeating part of the answer would give the whole answer.)

[The enthusiastic reader may be amused to see that

$$1/89 = .0112359\ldots71910112359\ldots$$

and that the digits 0, 1, 1, 2, 3, 5 are the first Fibonacci numbers.]

25. MORE SHIFTY MULTIPLES

Suppose our number is $N = abc \ldots z$ and it is multiplied by k to produce $kN = bc \ldots za$. Let $A = a$ and $B = bc \ldots z$, where B has n digits.

Then $N = A \times 10^n + B$ and $kN = 10\,B + A$. From these, we get $(k\,10^n - 1)A = (10 - k)B$. By trying $k = 1, 2, 3, \ldots$, we find that B is always greater than 10^n (i.e. it has at least $n+1$ digits), except in two cases.

When $k = 1$, we have lots of solutions such as $1 \times 3333 = 3333$. These are clearly too easy for Mr. Grind to set. When $k = 3$, then we have $(3 \times 10^n - 1)\,A = 7\,B$. If 7 divides A, then A must be 7 or 0, but 0 gives the trivial solution $A = B = 0$. If $A = 7$, then $B = 3 \times 10^n - 1$ has $n+1$ digits. But if 7 divides $3\times 10^n - 1$, then B has n digits when $A = 1$ or 2. Some elementary number theory is needed to see that this requires n to have the form $6m - 1$, but one can easily discover the first case, $m = 1$, $n = 5$, $B = 42857\,A$. For $A = 1$, we have $3 \times 142857 = 428571$, while $A = 2$ gives $3 \times 285714 = 857142$. The next case is when $n = 11$ and we get $3 \times 142857142857 = 428571428571$, etc.

26. SUM MAGIC

The missing digit is easily determined since $G + H + I = 18$. To see this, consider the sum S of all the digits: $S = A+B+C+D+E+F+G+H+I$, which must be equal to $1+2+3+4+5+6+7+8+9 = 45$. Suppose the sum $ABC + DEF$ has no carries. Then $A + D = G$, $B + E = H$ and $C + F = I$, so $S = 2(G + H + I) = 45$, which is impossible. Suppose there is a carry in our sum, say from the units place to the tens place. Then $C + F = 10 + I$, $B + E + 1 = H$, $A + D = G$, from which $S = 2(G + H + I) + 9 = 45$ and $G + H + I = 18$. The same result holds if there is one carry from the tens place to the hundreds place. If there are two carries, then we get $S = 2(G + H + I) + 18 = 45$, which is impossible.

I find 21 solutions of the sum, each of which has 16 forms, making 336 solutions in all. Since $A + D = D + A$, there are two ways to arrange the A and D. Similarly for B and E, C and F, and this gives 8 forms from one solution. Now consider a solution such as $271 + 683 = 954$ which has a carry from the tens place. We can move the units to the front to get $127 + 368 = 495$ which has the carry from the units place. This gives two forms from one solution, making a total of 16 forms from each solution.

27. DIGITAL DIFFICULTY

For a number N to have n digits, it must satisfy $10^{n-1} \leq N < 10^n$. Letting $N = a^n$ and taking common logarithms, this gives

$$n - 1 \leq n \ \log a < n.$$

The right-hand inequality tells us that $a < 10$ and the left side leads to $n \leq 1/(1 - \log\ a)$. For $a = 1, 2, \ldots, 9$, there are 1, 1, 1, 2, 3, 4, 6, 10, 21 values of n, giving 49 values in all. The largest is $9^{21} =$ 1 09418 98913 15123 59209.

28. ON THE SQUARE?

Finding all the solutions is a bit tedious to do by hand, so I let my computer do it. It found 20 solutions, given below. The underlined solutions were found by the previous author.

$$001, 010, 020, 050, 081, 100, 112, \underline{162}, 200, \underline{243},$$
$$\underline{324}, \underline{392}, 400, \underline{405}, 500, \underline{512}, \underline{605}, \underline{648}, \underline{810}, \underline{972}.$$

The previous author has omitted the first five solutions as not being really three-digit numbers. Then I thought he was omitting numbers with zeroes, but the presence of 405 in his list and the absence of 112 showed this was not right. It seems that he only considered cases where the first digit is not zero and the three digits are different!

29. SUMS OF POWERS OF DIGITS

There doesn't seem to be any simple way to find the PDIs with $k = 3$ other than making a table of the cubes of the digits and a lot of trial and error, though some limitations can be deduced from the following.

In general, for a given k, we can see that the number of digits of a PDI cannot be arbitrarily large as follows. If a number N has n digits, then $10^{n-1} \leq N < 10^n$. But the sum of the k-th powers of the digits of N is maximal when all the digits are 9, so this sum is $\leq n9^k$. When $n9^k < 10^{n-1}$, i.e. $9^k < 10^{n-1}/n$, then an n-digit number N cannot be a PDI for k-th powers. Since the right-hand side of this last inequality is increasing with n, there are only a finite number of values of n which work for a given k.

For $k = 3$, this shows that $n < 5$. Since 4×9^3 is 2916, a 4-digit solution will have to contain a 1 or a 2 and a little more work shows that there are no 4-digit solutions. 1-digit and 2-digit solutions are pretty easily disposed of, but it takes hard work to find the four 3-digit solutions are: 153, 370, 371, 407.

When $k = n$, our general argument shows that we cannot have a PPDI if $n9^n < 10^{n-1}$. It is not hard to see that the right-hand side of this last inequality grows faster than the left-hand side and first exceeds the left-hand side for $n = 61$, so there are only a finite number of PPDIs.

[Since writing this, I have found that the PPDIs were reported in 1981 and 1993. There are 88 of them (including 1, but excluding 0.]

30. EVEN MORE SHIFTY MULTIPLES

One can use the methods used before, but there is a sneakier method which gets the answers with less work. First consider moving the last digit to the front and let the number have decimal digits $a, b, \ldots, c, d$, so we have (3/2) $ab \ldots cd = dab \ldots c$ or $3ab \ldots cd = 2dab \ldots c$. Let A be the repeating decimal $.ab \ldots cdab \ldots cdab \ldots$ and let B be the repeating decimal $.dab \ldots cdab \ldots cdab \ldots$.

Then $3A = 2B$. But $10B = d.ab \ldots cdab \ldots = d + A$, hence $15A = 10B = d + A$, so $A = d/14$.

Because 14 is even, odd values of d do not give a purely periodic decimal expansion. Hence A is a multiple of 1/7, so our desired number is a multiple of 142857 and it must be an even multiple less than 666667 so that one can take 3/2 of it.

This gives just two solutions: (3/2) 285714 = 428671 and (3/2) 571428 = 857142.

[Of course we have longer solutions such as (3/2) 285714285714 = 428571428571, but these are not really different.]

Now consider moving the first digit to the end. Proceeding as above, we have $3\ ab \ldots cd = 2\ b \ldots cda$, $A = .ab \ldots cdab \ldots$, $B = .b \ldots cdab \ldots$, $3A = 2B$, $10A = a + B$, $20A = 2a + 2B = 2a + 3A$, so $A = 2a/17$. Since 1/17 is a purely periodic decimal, we get solutions for $a = 1, 2, 3, 4, 5$. When $a = 2$, the solution is 1 17647 05882 35294 or 1,176,470,588,235,294.

[Alternatively, the second problem is the same as the first with multiplier 2/3.]

31. MIXED-UP MULTIPLICATION

The problem is asking for $ab \times cd = ba \times dc$. The decimal number ab has the value $10a+b$, so the problem wants

$$(10a+b)(10c+d) = (10b+a)(10d+c),$$

which is

$$100ac + 10ad + 10bc + bd = 100bd + 10bc + 10ad + ac,$$

in which the middle terms cancel, leaving us with $99ac = 99bd$ or $ac = bd$. The cases $a = b, c = d$ and $a = d, c = b$ give only trivial solutions, so we really want genuinely different products ac and bd. Checking a times table, we find only 9 examples:

$$\begin{array}{lll} 1 \times 4 = 2 \times 2; & 1 \times 6 = 2 \times 3; & 1 \times 8 = 2 \times 4; \\ 1 \times 9 = 3 \times 3; & 2 \times 6 = 3 \times 4; & 2 \times 8 = 4 \times 4; \\ 2 \times 9 = 3 \times 6; & 3 \times 8 = 4 \times 6; & 4 \times 9 = 6 \times 6. \end{array}$$

From each of these, we can generally find two solutions — e.g. from $1 \times 6 = 2 \times 3$, we get $12 \times 63 = 21 \times 36$ and $13 \times 62 = 31 \times 26$. However, when one product has two equal factors, then only one solution occurs. This happens four times, so we get 14 solutions in all.

Perelman assumed that his digits are all positive. If we allow zeroes, we get some more solutions of the degenerate forms: $00 \times cd = 00 \times dc$ and $0b \times c0 = b0 \times 0c$.

More generally, one wants $ab \times cd$ to be a product of two two-digit numbers with the same digits. There are $4! = 24$ ways to permute the digits, but we can make the first factor contain a to reduce to 12 cases. After six pages of careful analysis, I realized I was neglecting some cases, so I let my computer find solutions. At first, I got too many trivial cases and it took a little care to suppress these. I found four new solutions, each of which has two forms:

$$01 \times 64 = 04 \times 16; \quad 01 \times 95 = 05 \times 19;$$
$$02 \times 65 = 05 \times 26; \quad 04 \times 98 = 08 \times 49.$$

The other forms are: $10 \times 64 = 40 \times 16$, etc., which do not have leading zeroes.

32. PERMUTED PRODUCTS

The answers are: $3 \times 51 = 153$; $6 \times 21 = 126$; $8 \times 86 = 688$.

33. RUBBED OUT!

After some trials, one notes that the units column must add up to an amount $U = 1, 11$ or 21. Depending on the amount carried, the tens column must add up to $T = 1, 11, 21$ or $0, 10, 20$ or $9, 19$, and the same holds for the hundreds column, whose total is H. These sums can be achieved in a limited number of ways.

$$1 = 1; \quad 11 = 1 + 3 + 7; \quad 21 = 5 + 7 + 9;$$
$$0 = 0; \quad 10 = 1 + 9 \text{ or } 3 + 7; \quad 20 = 1 + 3 + 7 + 9;$$
$$9 = 9 \text{ or } 1 + 3 + 5; \quad 19 = 3 + 7 + 9.$$

Because of the carrying, it is easiest to consider cases by looking at U first, then T, and then H is determined, but I will write the sums down in the usual order — H, T, U. The first case is then $H, T, U = 11, 1, 1$ and there is only one way these can occur, containing 137 in the hundreds column, 1 in the tens column and 1 in the units column, giving the sum $111 + 300 + 700$. This sum has 5 non-zero digits present, hence 10 digits are rubbed out. There are 7 cases which can occur, but some of them have multiple possibilities for the sum and there are 18 solutions altogether. Below I list the cases and the number of solutions and the number of digits present and rubbed out in each case. So we can have 5 through 10 digits present, or indeed rubbed out. For 5, 6, ... , 10 digits rubbed out, there are 1, 1, 3, 6, 5, 2 solutions, so the cases of 5 and 6 rubouts are probably the hardest to find.

H	*T*	*U*	Number of solutions	Digits present/rubbed out	
11	1	1	1	5	10
10	11	1	2	6	9
9	21	1	2	5,7	10,8
11	0	11	1	6	9
10	10	11	4	7	8
9	20	11	2	8,10	7,5
10	9	21	4	6,6,8,8	9,9,7,7
9	19	21	2	7,9	8,6

34. NEW CENTURY COMING UP

Since 100 leaves a remainder 1 after casting out 9s, a pure addition to make 100 must have the digits $1, 2, \ldots, n$ adding to a total which leaves 1 after casting out 9s. This is seen to happen with $n = 1, 4, 7$. The cases $n = 1$ and 4 seem to require exceptional operations such as logarithms, square roots, etc., in order to produce 100. But $n = 7$ gives us two nice solutions: $1 + 2 + 34 + 56 + 7$ and $1 + 23 + 4 + 5 + 67$. There are a number of less nice but purely additive solutions. Since one can reposition the tens digits in many ways (e.g. $1 + 23 = 21 + 3$), the solutions are best described by seeing that the units digits have to add up to 20 and there are five sets of digits which do this: 12467, 13457, 23456, 3467, 2567.

For other values of n, here are some reasonable solutions, but they get more difficult and complex as n decreases.

$$
\begin{array}{ll}
n = 8 & 12 + 3 - 4 + 5 + 6 + 78 \\
n = 6 & -1 + 2 + 34 + 65; \ -1 + 2 + 43 + 56 \\
n = 5 & 5(2^4 + 1 + 3) \\
n = 4 & [3/(.4 - .1)]^2
\end{array}
$$

35. THE RATIO OF A NUMBER TO THE SUM OF ITS DIGITS

Let a two-digit number be ab, so its value is $N = 10a + b$ and its digital sum is $S = a + b$. Now S divides N if and only if S divides $N - S = 9a$. If $b = 0$, this clearly holds and we have 9 solutions: $10, 20, \ldots, 90$. If $b > 0$, then $a + b$ is bigger than a, so it must have some factor in common with 9, i.e. it must be a multiple of 3. So we examine the cases.

$a+b=3$	gives $N = 12$ or 21.
$a+b=6$	requires a to be even, so we get $N = 24$ or 42.
$a+b=9$	gives $N = 18, 27, \ldots, 81, 90$.
$a+b=12$	requires a to be a multiple of 4, so we get $N = 48$ or 84.
$a+b=15$	requires a to be a multiple of 5 and there are no solutions.
$a+b=18$	requires a to be even and there are no solutions.

All together, there are 23 solutions.

The only examples where N is also divisible by the product of its digits are 12, 24, 36.

The only examples where $N = SP$ are 0, 1, 135, 144.

36. A DIGITAL CURIOSITY

For this sort of problem, there is often some way to do the problem by hand, but finding a manual method is often more work than writing a simple computer program. From $(AB + CD)^2 = ABCD$, we can deduce that $D = 0, 1, 4, 5, 6, 9$. For each value of D, there are at most five values of C that can occur. Further, for each value of D, there are at most two values of B such that $(B + D)^2$ ends in D. This rather reduces the work, but it is still a bit tedious to check through all the cases to find the following five solutions.

$$(00+00)^2 = 0000;\ (00+01)^2 = 0001;\ (20+25)^2 = 2025;$$
$$(30+25)^2 = 3025;\ (98+01)^2 = 9801.$$

37. TWO AND TWO ARE FOUR, OR FIVE

Perhaps the most natural way to start on such a problem is to make a table of the possible successors of each digit. I'll assume you've done this and have it to examine. Observe that every digit has two successors, but also has two predecessors. E.g., 6 can be preceded by 3 or 8; 9 can be preceded by 4 or 9.

If we have a cycle of the ten digits, then 0 must occur somewhere in it. Now 0 has successors 0 and 1, and taking 0 leads to a cycle of length 1. Further 0 has predecessors 0 and 5 and again we can't use 0. Hence our cycle must contain the segment 5, 0, 1. Looking at 9, we see we must have the segment 4, 9, 8. This leaves only a few cases to try and we readily find three solutions:

$$\ldots, 5, 0, 1, 2, 4, 9, 8, 6, 3, 7, 5, 0, \ldots;$$
$$\ldots, 5, 0, 1, 3, 6, 2, 4, 9, 8, 7, 5, 0, \ldots;$$
$$\ldots, 5, 0, 1, 3, 7, 4, 9, 8, 6, 2, 5, 0, \ldots.$$

[Adapted from a booklet of a lecture by Fred. Schuh, in Dutch, in 1944.]

38. PRIME PAIRING

There are 20 two-digit primes: 11, 13, 17, 19, 23, 29, 31, 37, 41, 43, 47, 53, 59, 61, 67, 71, 73, 79, 83, 89, 97. No two-digit prime can end in 2, 4, 5, 6 or 8, so the primes starting with those digits can at most occur in one place, namely at the start of the sequence. The 10 other primes can be arranged into a sequence of 11 digits in several ways, e.g. 19737131179. This has maximal length as all 10 other primes are used. We can precede this by a 4 or a 6 to get a sequence of 12 digits of the desired form and this is the maximal length.

The number of solutions is somewhat more than I anticipated and I had to use a computer. But we can simplify a bit. First we ignore the leading 4 or 6, so we are looking for a sequence of digits 1, 3, 7, 9 with the desired property. We can view this as a graph on these four points with an edge going from a to b if and only if ab is prime. What we want is what is called an Euler path in this graph. Examination shows that there is one more edge from 1 than to it, and one more edge into 9 than from it, while 3 and 7 have as many incoming edges as outgoing ones. Hence an Euler path must start at 1 and end at 9. We can simplify a bit further, as the edge 11 can be left out and then later re-inserted at any of the three places where 1 will occur. With all these simplifications, it is still a bit tedious to find all paths by hand — my computer finds 32 of them. These can each have 11 inserted in three places, giving 96 paths using 11. Adding 4 or 6 at the beginning gives us 192 solutions of the maximal length 12.

Chapter 3

39. SUM SQUARED

Jessica should take the bet as Henrietta loses. Consider the four numbers in the upper left of the array, and call them the basic values a, b, c, d. The array then looks like this.

$$\begin{array}{ccccc} a & + & b & = & a+b \\ + & & + & & + \\ c & + & d & = & c+d \\ = & & = & & = \\ a+c & + & b+d & = & a+b+c+d \end{array}$$

Now the least value that $a+b+c+d$ can have is $1+2+3+4=10$, so this entry cannot be a single digit value. One might try to get solutions for Henrietta by allowing one of a, b, c, d to be 0, but then the results of the additions would not be different values.

Looking for solutions like Jessica's example, we see that we can interchange the first two rows and/or the first two columns and/or reflect in the diagonals to make a be the smallest of the basic values and also make $b < c$. By hand trial and error, I find only three solutions, without any carry, each of which has 8 forms.

$$\begin{array}{ccccc} 1 & + & 3 & = & 4 \\ + & & + & & + \\ 6 & + & 2 & = & 8 \\ = & & = & & = \\ 7 & + & 5 & = & 12 \end{array} \qquad \begin{array}{ccccc} 1 & + & 3 & = & 4 \\ + & & + & & + \\ 7 & + & 2 & = & 9 \\ = & & = & & = \\ 8 & + & 5 & = & 13 \end{array} \qquad \begin{array}{ccccc} 1 & + & 4 & = & 5 \\ + & & + & & + \\ 7 & + & 2 & = & 9 \\ = & & = & & = \\ 8 & + & 6 & = & 14 \end{array}$$

If we allow a carry in the addition of $b+d$, we get some more solutions. We might hope to make Henrietta happy if the bottom row added up to a single digit number which was distinct from the others, but a bit more searching (which I let my computer do) reveals only the following five sets of values for a, b, c, d: $1, 7, 2, 3$; $1, 7, 4, 5$; $1, 8, 2, 5$; $1, 8, 3, 4$; $2, 6, 4, 5$. In no case do we get all nine digits occurring. (Here, each solution has only two forms, obtained by interchanging the first two rows.)

40. USING ALL THE DIGITS

Let the three numbers be *ABC, DEF, GHI*. Since $3 \times ABC$ is less than 1000, we know $ABC \leq 333$, so $A = 0, 1, 2$ or 3. If A is 0, then D must be 1 and G must be 2 and this forces B to be at least 6.

We know that F is the units digit of $2C$ and I is the units digit of $3C$. The possible triples *CFI* are: 000, 123, 246, 369, 482, 505, 628, 741, 864, 987. Clearly C cannot be 0 or 5. Also $C = 1$ makes $F = 2$, $I = 3$ which forces $A = 0$ and then we want $D = 1$, but 1 is already used. Hence C cannot be 1.

If C is 2, then the triple *BEH* must be one of the above triples and the only one not using 2, 4 or 6 is 987. Trying $A = 1$ and 3, we find one solution: 192, 384, 576.

If $C = 3$, then the triple *BEH* is again one of the above and the only possibilities are 482 and 741. Only the second gives a solution: 273, 546, 819.

If $C = 4$, then the triple *CFI* is 482 and there is a carry of 1 into H. Thus the possible triples *BEH* are the above triples with the last digit increased by one, namely: 001, 124, 247, 360, 483, 506, 629, 742, 865, 988. Only 360 and 506 are possible, but in both cases there is no solution.

If $C = 6$, then the triple *CFI* is 628 and the triples *BEH* must be our original triples with one added to the last two digits, namely: 011, 134, 257, 370, 493, 516, 639, 752, 875, 998. The only possible *BEH* are 134, 370, 493 and none of these gives a solution.

If $C = 7$, then $CFI = 741$ and the triples *BEH* must come from 012, 135, 258, 671, 494, 517, 630, 753, 876, 999. Only 258 and 630 are possible and both give solutions: 327, 654, 981 and 267, 534, 801.

If $C = 8$, then $CFI = 864$ and *BEH* must be one of the same triples as for $C = 7$. The possibilities are 012, 135, 517, 753, but only the last gives a solution: 078, 156, 234.

If $C = 9$, then $CFI = 987$ and BEH must be one of the triples from $C = 7$. The only possible cases are 012, 135, 630, but only the second gives a solution: 219, 438, 657.

41. A MAGICAL CROSS

First let's see if there can be any other magic sums. Let the number in the intersection be C (for cross) and let the sum be S (for sum). When we add up the row and the column, we have $2S$ and this consists of all the numbers from 1 to 9 with C counted twice. Since $1 + 2 + \cdots + 9 = 45$, this tells us that: $2S = 45 + C$. Hence, C must be an odd number: 1, 3, 5, 7, 9 and the corresponding S values are 23, 24, 25, 26, 27. For each case, one can determine all the solutions fairly easily if one ignores the obvious re-arrangements — one has just to find four of the numbers which add up to $S - C$, as the remaining four numbers must also add up to $S - C$. We can save 50% of the searching by just looking for four numbers which include the smallest available number. We can save 40% of the work by noting that the solutions for $C = 1, 3$ correspond to the solutions for $C = 7, 9$ by replacing every number x by $10 - x$. The solutions are determined by the following rows of four numbers.

$C = 1$,	$S = 23$	2389,	2479,	2569,	2578
$C = 3$,	$S = 24$	1479,	1569,	1578	
$C = 5$,	$S = 25$	1289,	1379,	1469,	1478

In the case $S = 25$, one of the solutions is not obtained by re-arranging pairs which add to 10, namely the solution given by 1478.

For each of these solutions, we can put the given numbers in the row or in the column, and we can re-arrange the row in $4! = 24$ ways and the column in 24 ways, giving 1152 obvious re-arrangements of each solution.

42. GOING ROUND IN CIRCLES

Let five consecutive numbers be A, B, C, D, E, followed by their opposites A', B', C', D', E'. Consider two adjacent numbers A, B and their opposites A', B'. Then we want $A + B = A' + B'$, which is the same as saying that $A - A' = B' - B$. This also works for the next adjacent pairs, so we must have $A - A' = B' - B = C - C' = D' - D = E - E' = (A')' - A' \ldots$ The last equality only holds because we have an odd number of pairs and hence there can be no solution when we try to arrange $1, \ldots, 8$ in this way. For our original case with 10 numbers, we see that we have to pair them into 5 pairs with the same difference d. With a little trial and error, one finds that the only ways this can be done are when the common difference, d, is 1 or 5 and this determines the pairs. Any arrangement of these pairs such that big ends alternate with little ends is a solution of the problem. For example, for difference 5, the pairs are 1,6; 2,7; 3,8; 4,9; 5,10, and a solution is: 1,7,3,9,5,6,2,8,4,10. For difference 1, a solution is: 1,4,5,8,9,2,3,6,7,10. If we fix the number 1 in place, there are $4! = 24$ solutions, but half are reflections of the other half, so there are 12 really distinct solutions for $d = 5$ and another 12 for $d = 1$, making 24 in all.

[We can determine the possible values of the common difference d as follows. The first d numbers can only be paired with the numbers that are d larger, leading to a pairing off of the first $2d$ numbers. For example if $d = 2$, we must have the pairs 1,3 and 2,4. Then the next $2d$ numbers must be similarly paired off, etc., so that $2d$ must be a divisor of 10, i.e. d must divide 5. More generally, if we have $2n$ numbers, where n is odd, then d must be a divisor of n, and each such divisor gives a pairing. Each such pairing gives $(n-1)!/2$ solutions.]

43. A MAGIC HOURGLASS

Since the numbers $A, B, \ldots, G$ are just the numbers $1, 2, \ldots, 7$ in some order, they all add up to $1 + 2 + \cdots + 7 = 28$, which we denote T. Let the magic constant, i.e. the sum of each the lines, be S. From the horizontals, we see that $2S + D = T$. From the lines through the center, we see that $3S = T + 2D$. These have the unique solution $S = 12$, $D = 4$. (So I could have given you the value S — but it's more fun to see there's only one possible value.) Hence $A+G = B+F = C+E = 8$, which can only hold if the pairs A, G; B, F; C, E are the pairs 1,7; 2,6; 3,5 in some order. One such order has $A = 1$, $G = 7$ and B, F is either 2,6 or 6,2, while C, E is either 3,5 or 5,3. But in order for $A + B + C = 12$, we have to take $B = 5$, $C = 6$. Thus the solution is essentially unique, but we can permute A, B, C in $3! = 6$ ways and we can interchange the top and bottom rows, so this solution has 12 different but equivalent patterns.

44. THE MAGIC OF 67

If a 4 by 4 magic square is formed from the 16 consecutive numbers: $a+1, a+2, \ldots, a+16$, then the sum of all the numbers, namely $8(2a+17)$, must be four times the magic constant C, so $C = 2(2a+17) = 4a+34$. Thus C must have a remainder of 2 when divided by 4 and so 67 can not be the constant for a consecutive magic square. When $a = 0$, there are many examples of such standard magic squares and we can add a to each value to increase the constant by $4a$, so we can have consecutive magic squares for all constants of the form $4a + 34$. If we only permit positive entries, we get the constant values 34, 38, 42, ... If we allow a zero entry, we can also get a constant of 30, but smaller values would require negative entries.

The technique for getting other constants is implicit in the square used by my neighbor. If we subtract 8 from each number we get the first square shown below. We see that this is almost a standard magic square except that the last four numbers, underlined below, have been increased by one. Looking closely, we see that these four numbers are so distributed that there is just one on each of the 10 lines which should add to the magic constant. Consequently, diminishing each of these by one will give us the standard magic square shown secondly below. In this square, each of the four groups 1–4, 5–8, 9–12 and 13–16 are located with just one from each group on each of the 10 magic lines. (Note that all such patterns are congruent.) Consequently, we can add one to the last 4 numbers to increase the constant by one, or we can add one to the last 8 numbers to increase the constant by two, or we can add one to the last 12 numbers to increase the constant by three. Each of these uses an almost consecutive set of integers. Adding one to all the numbers increases the constant by four, using a consecutive set of integers as already investigated above. Consequently, we can achieve any magic constant greater than or equal to 34 — we go up in steps of 4 and then increase by 1, 2 or 3. If our constant has remainder 2 when divided by 4, we can do it with consecutive numbers, otherwise we have to use almost consecutive numbers. If we permit a zero entry, then we can also have constants 30, 31, 32, 33. Smaller constants will require negative entries.

8	11	$\underline{15}$	1
$\underline{14}$	2	7	12
3	$\underline{17}$	9	6
10	5	4	$\underline{16}$

8	11	14	1
13	2	7	12
3	16	9	6
10	5	4	15

In 1995 my colleague Simon Nightingale told me that there is an almost consecutive square with constant 33 in Barcelona on the west facade — the Passion Facade — of the Temple of the Sagrada Familia. I have since gone to see this. The Guide Book calls it a cryptogram, with constant 33 — Christ's age at the time of his passion. The entries are in the left square below:

1	14	14	4
11	7	6	9
8	10	10	5
13	2	3	15

1	14	$\underline{15}$	4
$\underline{12}$	7	6	9
8	$\underline{11}$	10	5
13	2	3	$\underline{16}$

This square has been adapted by subtracting one from the four underlined entries in the right-hand square, which is is one form of the oldest and most common 4×4 magic square — a 180° rotation of it is in Dürer's famous *Melencolia I*. However, I find the presence of repeated values unsatisfying, especially as one can give a square with no repeats and the numbers being almost consecutive, as described above, though this involves the use of a zero value, namely:

8	11	15	0
14	1	7	12
2	17	9	6
10	5	3	16

45. MAGIC TRIANGLES

In the pattern given, we want $A+B+D = A+C+F = D+E+F = S$, where S is the constant magic sum which needs to be determined. The numbers A to F comprise some arrangement of the numbers 1 to 6, so we have that $A + \cdots + F = 1 + \cdots + 6 = 21$. If we add the three edge sums together, we get all the numbers $A, \ldots, F$, with A, D, F repeated, so that we have $3S = 21 + A + D + F$. This forces $A + D + F$, which we denote as T, to be a multiple of three. The smallest value of T is $1+2+3=6$ and the largest value is $4+5+6=15$, giving us just four possible values of T: 6, 9, 12, 15. The values of A, D, F determine T and hence S and hence the values of B, C, E, so we just have to try all the ways of producing these values of T, and by symmetry of the figure, we can assume $A < D < F$.

For $T = 6$, we have $S = 9$ and the only possible values of A, D, F are 1, 2, 3, giving us the first triangle below.

For $T = 9$, we have $S = 10$ and the possible values of A, D, F are: 1, 2, 6; 1, 3, 5; 2, 3, 4. In the first case, we would need $B = 7$, and in the third case, we would need $C = 4$, while the second case works, giving us the second triangle below.

We can continue for $T = 12$ and $T = 15$, but there is a standard symmetry in these problems. If we replace every number X by $7 - X$, then the values of S and T are replaced by $21 - S$ and $21 - T$. Thus the solutions for $T = 6$ are symmetric to those for $T = 15$ and those for $T = 9$ are symmetric to those for $T = 12$. For completeness, I include the other solutions as the third and fourth triangles below, but in the forms symmetric to the first two.

```
    1           1           6           6
   6 5         6 4         1 3         1 2
  2 4 3       3 2 5       4 5 2       5 3 4
```

46. A DIFFERENT MAGIC TRIANGLE

After a bit of trying, one realizes that 6 cannot be on an edge, and we can assume it is on the top, i.e. we can assume $A = 6$. Then the values of D and F determine the other values and we can assume $D < F$. This gives just 10 pairs of values to examine and we soon notice that neither D nor F can be 3, which reduces us to just 6 cases. Systematically trying $D, F = 1, 2;\ 1, 4;\ 1, 5;\ 2, 4;\ 2, 5;\ 4, 5$, we find that only the second and fifth cases work, giving two solutions.

```
    6           6
   5 2         4 1
  1 3 4       2 3 5
```

Actually, we can also observe that neither B nor C can be 3, so that $E = 3$, which immediately reduces us to the two cases above. Because we are using differences, the usual symmetry argument doesn't apply, but there is a kind of symmetry between the two solutions.

47. MAGIC CIRCLES

The conditions of the problem tell us that we want $A + C + D + E = A + B + D + F = E + B + C + F = S$, where S is the magic sum for each circle. Adding these together, we get $2(A + B + C + D + E + F) = 3S$. But $A + B + C + D + E + F$ is some rearrangement of the digits, $1, 2, \ldots, 6$, so the sum $A + B + C + D + E + F = 1+2+3+4+5+6 = 21$ and hence $S = 14$ is the only possible magic constant. Now $A + B + D + F = 14$ implies $C + E = 7$, and symmetrically we have $A + D = B + F = 7$. But there are just three ways to get two of our digits to add up to 7, namely $1+6$; $2+5$; $3+4$. Disposing these three pairs of digits into the pairs of positions A, D; B, F; C, E gives us all solutions. There are $3! \times 2^3 = 48$ ways to do this. However, one might consider rotations of the pattern to give the same solution, in which case there are 16 distinct solutions. If one also considers reflections as giving the same solution, then there are 8 distinct solutions, which I enumerate:

$$ABCDEF = 123645, 124635, 153642, 154632, 623145,$$
$$624135, 653142, 654132.$$

48. ANOTHER MAGIC CIRCLE

Let the magic constant be S. There are 5 diameters which will add up to the magic constant S. Adding all these together will give us that $5S$ is the sum of all the first 11 numbers with the center value C counted 4 extra times. That is, $5S = 66 + 4C$. In order for $66 + 4C$ to be divisible by 5, we must have $C = 1, 6$ or 11, with $S = 14$, 18 or 22. In each case, it is easy to see that the ends of the diameters have to contain the highest and lowest available values, then the next highest and next lowest, etc. For example, for $C = 6$, the ends of the diameters must be 1, 11; 2, 10; 3, 9; 4, 8; 5, 7. These diameters can be arranged in a number of ways which are not significantly different. The 5 diameters can be permuted in $5! = 120$ ways and each one can be reversed, making another factor of $2^5 = 32$, giving 3840 different arrangements. If we agree that arrangements are the same if one can be rotated (or rotated and reflected) into the other, then we can divide our number by 10 (or 20) to get the number of distinct patterns as 384 (or 192). Note that the solution for $C = 1$ is dual to that for $C = 11$ by replacing every number i by $12 - i$.

49. A MAGIC FRAME

Let S be the magic constant. When we add up all the sums making S, we get all the numbers with A, D, F, I occurring twice. Now the sum of all the numbers is $1 + 2 + \cdots + 10 = 55$. Letting $A + D + F + I = T$, we have $4S = 55 + T$. Now T must be at least $1 + 2 + 3 + 4 = 10$ and at most $7 + 8 + 9 + 10 = 34$. Hence $4S$ must be at least 65 and at most 89, so $S = 17, 18, 19, 20, 21, 22$. I tried finding all ways of adding four values to produce the corresponding Ts, but this became tedious and I wrote a program to do it for me. This produced ten solutions! Actually it produced many more at first, but I made it check that the smallest corner value was at A and that $B < C$ and $H < G$. This reduces the number of solutions by a factor of 16. The solutions are the following.

1 4 7 6	2 3 8 5	1 4 5 9	1 5 7 6	1 5 6 7
9 10	10 9	10 7	8 9	8 9
8 3 5 2	6 1 7 4	8 2 6 3	10 2 3 4	10 2 4 3
3 1 10 5	2 3 6 9	3 2 5 10	6 3 5 8	6 1 5 10
9 8	8 7	9 6	7 4	7 4
7 2 4 6	10 1 5 4	8 1 7 4	9 1 2 10	9 2 3 8

50. A MAGIC TRIANGLE SQUARED

Let S be the magic constant, so that

$$S = A + B + C + D = D + E + F + G = G + H + I + J.$$

Set $T = A + D + G$. If we add the three sums which total S together, we will get all the integers $1 + 2 + \cdots + 9$ and the three corner values A, D, G, are repeated. Recalling that $1 + 2 + \cdots + 9 = 45$, we have $45 + T = 3S$. Now T is at least $1 + 2 + 3 = 6$ and at most $7 + 8 + 9 = 24$, so $51 \le 3S \le 69$ or $17 \le S \le 23$. For each value of S and the corresponding T, one can find all triples, A, D, G, which add to T and try filling in the rest of the triangle. I did it by hand, but felt I wasn't being careful enough, so I wrote a program and it found some errors in my handwork. There are 18 solutions, listed below as $A\ B\ C\ D\ E\ F\ G\ H\ I$.

1 5 9 2 4 8 3 6 7,	1 6 8 2 5 7 3 4 9,	1 5 9 4 2 6 7 3 8,
1 6 8 4 3 5 7 2 9,	2 5 9 3 1 8 7 4 6,	2 6 8 3 4 5 7 1 9,
1 6 8 5 2 4 9 3 7,	2 4 9 5 1 6 8 3 7,	2 6 7 5 3 4 8 1 9,
3 4 8 5 2 6 7 1 9,	4 2 9 5 1 8 6 3 7,	4 3 8 5 2 7 6 1 9,
3 4 8 6 1 5 9 2 7,	3 5 7 6 2 4 9 1 8,	3 2 9 7 1 5 8 4 6,
3 5 6 7 2 4 8 1 9,	7 2 6 8 1 5 9 3 4,	7 3 5 8 2 4 9 1 6.

Of these, just one has the additional property that the sum of the squares along the edges is also a constant, namely the eighth one above and the sum of the squares is 126.

51. YET ANOTHER DIFFERENCE TRIANGLE

The three entries in the bottom row determine the other three entries. Since 6 cannot be a difference, it must occur in the bottom row. We can reflect a solution in the vertical axis without essentially changing anything, so we can assume $D < F$. This still leaves 30 cases to consider, but most of them are immediately seen to be impossible and a few minutes' work reveals four solutions.

$$A, B, C, D, E, F = 3, 5, 2, 1, 6, 4; \quad 3, 4, 1, 2, 6, 5;$$
$$2, 3, 5, 4, 1, 6; \quad 1, 3, 4, 5, 2, 6.$$

52. STILL ANOTHER DIFFERENCE TRIANGLE

Let S be the 'magic sum', $S = A + D - B = A + F - C = D + F - E$. Adding all three of these gives $3S = 2(A + D + F) - (B + C + E)$. Adding and subtracting $A + D + F$ gives

$$3S = 3(A + D + F) - (A + B + C + D + E + F) = 3(A + D + F) - 21,$$

so $S = A + D + F - 7$. Now $6 = 1 + 2 + 3 \leq A + D + F \leq 4 + 5 + 6 = 15$, so $-1 \leq S \leq 8$. Because the problem can be rotated and reflected without changing the behavior, we can assume $A < D < F$. This gives 20 cases to examine, but we can't have A, D or F being equal to S as that would make two entries be equal. This reduces the problem to eight cases, each of which is a solution of the problem.

$ADFBCE$ = 123456; 124356; 135246; 145236; 236145; 246135; 356124; 456123.

If we replace each entry X by $7 - X$, we change the sum S to $7 - S$ and this converts the first four solutions to the second four solutions.

53. A DICEY MAGIC SQUARE

It is immediately obvious that one could rotate the given solution to three other solutions.

This has not been ruled out in the original problem, but we naturally will consider these four solutions as being the same. But it is also obvious that there is another solution with each entry being a three.

A	B	C
D	E	F
G	H	I

For reference, we label the entries of the array as above.

If 6 occurs in a solution, then the other two entries in any line through the 6 must be 1 and 2. If 6 occurs in a corner, say A, then C, E, G must all be 1 or 2 and these cannot add up to 9. If 6 occurs in the center, then every other entry must be 1 or 2 which is clearly impossible. If 6 occurs on an edge, say at B, then one of A and C must be 1. Suppose $A = 1$. Now E must be 1 or 2 and the only possibility is to have $E = 2$, $I = 6$, but we have already shown that 6 cannot occur in a corner. Hence 6 cannot occur in a solution.

Now consider 5 occurring. Then the other entries in a line through it must be 1, 2 or 3. If 5 occurs in a corner, say A, then C, E, G must all be 1, 2 or 3 and the only way they can add to 9 is if they are all 3 and the solution below is forced.

5	1	3
1	3	5
3	5	1

If 5 occurs in the center, then all other values must be 1, 2 or 3 and this would force them all to be 3. If 5 occurs on an edge, say at B, then one of A, C must be a 1 or a 2. Suppose $A = 1$. Then $E = 1, 2$ or 3 and

$A + E + I = 9$. $E = 1$ is too small. $E = 2$ forces $I = 6$ which has already been eliminated. $E = 3$ leads to $I = 5$ which is the case previously considered and gives a rotation of the previous solution. But we could have $A = 2$ and so $C = 2$. Again examining the possible cases for E, we find one solution, shown below.

2	5	2
3	3	3
4	1	4

Hence we need only look for solutions using 1, 2, 3, 4. Now the only way a 1 can occur is if the other entries in any line through it are both 4s, and it is easily seen that a 1 cannot occur at a corner, in the center or at an edge. This leaves us to consider solutions using just 2, 3, 4 and the only possible triples that can occur on a line are the different orders of 2, 3, 4 and 3, 3, 3. If all corners are 3, we get the solution with all 3s. Otherwise some corner must be a 2, say A. Trying the two possible top rows of 2, 3, 4 and 2, 4, 3, we find that only the latter gives a solution and it is a rotation of the solution given in the source.

So there are four distinct solutions. Three of them have four rotational forms, so if we fail to identify rotations, we would get 13 solutions. This is easily checked with a computer. Readers may like to consider letting the range of values be $1, 2, \ldots, N$ and/or taking a different sum for each line. With a moderate bit of algebraic manipulation, one can show that E is one third of the sum for each line, which would have simplified the arguments above.

54. ANOTHER MAGIC FRAME

The problem wants us to use the digits $1, 2, \ldots, 8$ in the figure below so that each of the edge triples has the same sum, which I denote S.

A	B	C
H		D
G	F	E

That is, we want $A+B+C = S = C+D+E = E+F+G = G+H+A$. Note that $A + B + \cdots + H = 1 + 2 + \cdots + 8 = 36$. Considering the rows, we have $2S + D + H = 36$ and similarly, $2S + B + F = 36$, so that $B + F = D + H$. *A priori*, we could have $B + F$ as small as $1 + 2 = 3$, but then there is no way to find D, H with $D + H = 3$. In fact, the smallest possible value of $B + F = D + H$ is $5 = 1 + 4 = 2 + 3$. Similarly, the largest value is $13 = 8 + 5 = 6 + 7$. Hence $23 \leq 2S \leq 31$, and since $2S$ is even, we have $24 \leq 2S \leq 30$, and S can just take on the values 12, 13, 14, 15. Replacing each number x by $9 - x$ transforms S to $27 - S$, so we need only look at $S = 12, 13$, for which $B + F = D + H = 12, 10$.

There are only a few possible triples which add up to S in either case.

12 is obtained from $5, 4, 3$; $6, 5, 1$; $6, 4, 2$; $7, 4, 1$; $7, 3, 2$; $8, 3, 1$.
13 is obtained from $6, 5, 2$; $6, 4, 3$; $7, 5, 1$; $7, 4, 2$; $8, 4, 1$; $8, 3, 2$.

When $S = 12$, $B + F = D + H = 12$, and the only solution for this is $8 + 4 = 7 + 5$, and these then must be in the centers of the edges of our magic figure. The first and fourth of the possible triples cannot then occur in our magic figure as they use two of these edge values. This leaves just enough edge triples and there is essentially one solution:

$$A, B, \ldots, H = 1, 8, 4, 3, 6, 2, 5, 7.$$

When $S = 13$, $B + F = D + H = 10$, and there are three possible pairs: $8 + 2$, $7 + 3$, $6 + 4$. Now the value 1 doesn't occur in these pairs, so it cannot be in the center of an edge, i.e. it must be in a corner. Similarly, the value 5 must be in a corner, so we must have 1, 7, 5 along one edge and the other edge through 1 must contain 4 and 8. The two orders of 4 and 8 give us two solutions: 1, 7, 5, 6, 2, 3, 8, 4 and 1, 7, 5, 2, 6, 3, 4, 8.

Taking account of the two and one solutions for $S = 14, 15$, we have a total of six solutions — though each of these can be rotated and reflected to eight forms.

55. A GENERALIZED MAGIC SQUARE

I began by attacking T which is an odd number expressible as a sum of three distinct digits in at least two ways. The smallest sum of three distinct digits is $6 = 1 + 2 + 3$. But since T is odd, we start looking at $T = 7$. This is expressible as a sum of three distinct digits in only one way: $1 + 2 + 4$, so $T = 7$ can be eliminated.

Now consider $T = 9 = 6 + 2 + 1 = 5 + 3 + 1 = 4 + 3 + 2$. Then $S = 18 = 9 + 8 + 1 = 9 + 7 + 2 = 9 + 6 + 3 = 9 + 5 + 4 = 8 + 7 + 3 =$

$8 + 6 + 4 = 7 + 6 + 5$. The value of E must occur twice in the sums for T and so must be 1, 2, or 3. But E also must occur twice in the sums for 18 and neither 1 nor 2 does so. Hence $E = 3$ and, up to symmetry, there is just one way to use the sums $5 + 3 + 1$ and $4 + 3 + 2$, as below.

A	1	C
2	3	4
G	5	I

Now there is just one sum for 18 which contains a 1, so the top edge must be 8, 1, 9 or 9, 1, 8. But 8 and 2 do not occur together in any of the sums for 18, so we have $A = 9$, which gives us $C = 5$, $G = 7$, $I = 6$ and the solution is unique (up to symmetry of the figure).

Similar, but lengthier, reasoning applies for $T = 11, 13, 15$. The case $T = S = 15$ is the usual magic square and is known to have a unique solution, again up to symmetry. If one replaces each digit X by $10 - X$, one gets a solution with T, S replaced by $30 - T$, $30 - S$, so that the cases $T = 17, 19, 21, 23$ are dual to the cases $T = 13, 11, 9, 7$ and we do not need to examine them. I was surprised to find that there are no solutions for $T = 11$ or 13. I originally found one, but knowing that it is easy to make mistakes in such problems, I wrote a small program to check my work. It failed to find my second solution and checking showed that I had indeed made a mistake!

Perhaps the German author had done all this work, but I am surprised not to have ever seen this problem before as it has such a neat answer.

56. THE MAGIC OF SEVEN

We have $A + B + C = E + F + G = S$. So $2S + D$ contains all the seven numbers and must equal $1 + 2 + \cdots + 7 = 28$.

We also have $A + D + G = E + D + C = B + D + F = S$. This gives $3S = 28 + 2D$. Combining this with $2S + D = 28$, we find that $S = 12,\ D = 4$.

Now $A + E = 12$ has only one possible solution $7 + 5$ or $5 + 7$. By symmetry, let us assume $A = 7, E = 5$. Then all the values are determined: $A, B, C, D, E, F, G = 7, 2, 3, 4, 5, 6, 1$. So there is just one solution, together with its reflection in the horizontal mid-line.

Chapter 4

57. FUNNY MONEY

Let A, B, C be the amounts of pounds, shilling and pence. Then the value, in pence, is $V = 240A + 12B + C$. If I simply interchanged two values, say A and B, then the confused value would be $W = 240B + 12A + C$. If $V = W$, then we get $228A = 228B$, so $A = B$, which isn't really confusing. Similar arguments apply for exchanging A and C or exchanging B and C. Hence I must have written B, C, A or C, A, B instead of A, B, C, i.e. I shifted the figures forward or backward. The latter case is just the reverse of the former case, so we need only consider the former case. Then my confused value W is $240B + 12C + A$. Setting $V = W$, we get (*) $239A = 228B + 11C$, and we want A, B, C all non-negative and unequal. The equation (*) has an obvious solution: $A = B = C$, for any A. This gives us

$$239A = 228A + 11A.$$

Subtracting (*) from this leaves $0 = 228(A - B) + 11(A - C)$. Since 11 and 228 have no common factor, this can happen only if 11 divides $A - B$ and 228 divides $A - C$. Letting $K = (A - B)/11 = -(A - C)/228$, we have $B = A - 11K$, $C = A + 228K$. The first time that A, B, C can be unequal non-negative integers is when $A = 11$ and then we can take $K = 1$ to give $A, B, C = 11, 0, 239$, whose value is 11£ 19s 11d, which is the same as 0, 239, 11 and this is the least solution. The fact that I was thinking of 239d at all simply shows how confused I was.

[We can add one to each of A, B, C, getting solutions 12, 1, 240; 13, 2, 241; etc. until $A = 21$. At $A = 22$, there is another solution, for $K = 2$, so we have two solutions: 22, 11, 250 and 22, 0, 478. When A gets as large as 228, we can also take $K = -1$. In general, K can take on all the non-zero integer values between $-A/228$ and $A/11$, inclusive. (If one lets $[X]$ denote the integer part of X, then the number of solutions for a given A is $[A/11] + [A/228]$.)]

58. NO CHANGE

The maximum amount of coins less than a pound that Sarah could have without being able to make change for a pound is £1.43. She would have: one 50p, four 20p, one 5p and four 2p. With this, she can't even make change for any coin!

[In the US, one can have $1.19: one 50¢, one 25¢, four 10¢ and four 1¢, or one can have one 25¢, nine 10¢ and four 1¢. In the first case, one cannot make change for any amount, but in the latter case one can make change for a 50¢ piece.

In Canada, one can have the same amount, but with three 25¢, four 10¢ and four 1¢, or with one 25¢, nine 10¢ and four 1¢. In neither case can one make change for any amount.]

59. ALL CHANGE!

Since Sarah has given at least one of each coin, she has already given $50 + 20 + 10 + 5 + 2 + 1 = 88$p, leaving only 12p unaccounted for. This 12p can be given in any way, but obviously only can use 1, 2, 5 and 10 pence coins. We denote a way by use of exponents, so that $5^1 2^2 1^3$ denotes $5 + 2 + 2 + 1 + 1 + 1$. Then there are just 15 solutions:

$$10^1 2^1,\ 10^1 1^2,$$
$$5^2 2^1,\ 5^2 1^2,\ 5^1 2^3 1^1,\ 5^1 2^2 1^3,\ 5^1 2^1 1^5,\ 5^1 1^7,$$
$$2^6,\ 2^5 1^2,\ 2^4 1^4,\ 2^3 1^6,\ 2^2 1^8,\ 2^1 1^{10},$$
$$1^{12}.$$

If Sarah had enough money to give any one of these, she would have had to have the 88p plus one 10p, two 5p, six 2p and twelve 1p, making a total of £1.32.

Computing the total number of ways of making change for a pound is rather more tedious and is best left to a computer. My computer found 4562 ways. Needless to say, I haven't bothered to print these out. One can add an extra way if one counts using one £1 coin, which might be appropriate if one asked for change of a £1 note — except there aren't any £1 notes in the UK any more.

[For the US, the reasoning shows that Sarah must have already given $50 + 25 + 10 + 5 + 1 = 91$ cents, leaving only 9 cents to be variable. There are only two ways to do this: 9 single cents, or 4 cents and a 5-cent piece. The total amount to permit both ways is just $1.05. There are 292 ways to change a $.

For Canada, Sarah must have already given $25+10+5+1 = 41$ cents, leaving 59. There are 60 solutions of this and one needs $2.20 to be able to do all of these. Of these solutions, 6 use at least two of each coin. There are 60 ways to change a Canadian $.]

60. NO CHANGE!

There are actually several possible ways this can happen. Perhaps the easiest is that I owed my friend £2 and he only had £2 coins on him. Or I could owe him £4.99 and he has no 1p coins. Or I could owe him £4.90 and he has only 20p and 50p coins. There are many variations on these basic answers.

It can happen in America with $2 bills which one occasionally gets. The $4.99 case doesn't work as the US doesn't have 2¢ coins. Because the US has 25¢ coins, the last situation changes to owing $4.95 to someone who has only 10¢ and 25¢ coins.

(I found this idea in a 1935 American book where a man had to pay 5¢ and his friend couldn't change a $1 bill, but could change a $5 bill. I couldn't solve this and had to look at the answer which used a $2.50 gold piece and a $2 bill! (I've just found this also in a 1916 book.) One can create more examples from old English currency. Indeed a common problem in 17th to 19th-century textbooks involved my wanting to pay a friend £100 when I only had guineas ($= 21s =$ £1.05) and my friend only has pistoles (worth $17s =$ £0.85)!)

61. AN OLD MONEY PROBLEM

Let the number of pounds be P and the number of shillings be S, with $S < 20$. Then the amount £P Ss is equal to $20P+S$ shillings. Interchanging the numbers gives us £S Ps which is equal to $20S+P$ shillings and we want this to be equal to $40P+2S$. This gives us $18S = 39P$, hence $6S = 13P$. $S = P = 0$ is a rather trivial solution which I hadn't ruled out. The only solution with $0 < S < 20$ is $S = 13$, $P = 6$, i.e. £6 13s.

To see why I had to use old money, suppose we have a monetary system with base B. Then our desired relationship is $SB + P = 2PB + 2S$, so

$(B-2)S = (2B-1)P$. One can work from the latter equation, but it uses a little number theory, so I will just observe that this tells us that $P \leq S$. Now consider the small part, $2S$, of $SB + P = 2PB + 2S$. If $2S < B$, then the expressions have units parts P and $2S$ which must be equal. But this leads to $SB = 4SB$ which can only hold if $S = 0$ and then $P = 0$. So we must have $2S \geq B$, while $S < B$ gives us $B \leq 2S < 2B$, i.e. there is only a single carry from the units column. Hence $P = 2S - B$, $2P + 1 = S$. Combining these gives us $3P = B - 2$ and so we can solve for P if and only if $B - 2$ is divisible by three. $B = 20$ works, but $B = 100$ does not! (Solving for S gives $3S = 2B - 1$ and it is easy to see that one can solve for S if and only if one can solve for P.)

62. A PROFITABLE ERROR

Suppose Jessica's cheque was for A pounds and B pence. Clearly $99 \geq B \geq 0$, and in order for confusion to occur, we must also have $99 \geq A \geq 0$. Since Jessica made a profit, we also have $B > A$. Reducing everything to pence, Jessica's cheque was for $C = 100A + B$ pence and she received the reversed amount $R = 100B + A$. Her profit P is $R - C = 99(B - A)$. Since $P > 0$, i.e. $B > A$, the minimum profit is 99 pence, which occurs when $B = A + 1$. There are 99 cases, starting with $A = 0, B = 1, C = £00.01, R = £01.00$ and continuing to $A = 98, B = 99, C = £98.99, R = £99.98$. The largest profit is $99 \times 99 = 9801$ pence, which occurs when $B = A + 99$ and there is just one case: $A = 0$, $B = 99$, $C = £00.99$, $R = £99.00$. Of course, this last is not a realistic cheque, but $C = £1.99$, $R = £99.01$ gives a profit of £97.02.

63. A PROFITABLE RATIO OF ERROR

As before, let Jessica's cheque be for A pounds and B pence, i.e. for $C = 100A + B$ pence. The reversed amount is $R = 100B + A$ pence. The ratio of these is $R/C = (100B + A)/(100A + B) = (100r + 1)/(100 + r)$, where $r = B/A$ is given as greater than one. A little examination shows that this ratio increases with r, and this can be verified by a little manipulation of the expression. So the minimum ratio greater than one occurs when r is as small as possible. Since $99 \geq B > A \geq 0$, the minimum value of r is

99/98, giving the minimum ratio greater than one as

$$9998/9899 = 1.0100010102\,030508\ldots$$

for $A = 98, B = 99, C = £98.99, R = £99.98$. If one is happy with infinities, one can see that the maximum of R/C occurs when r is infinite, i.e. $A = 0$ and then $R/C = 100$. Alternatively, one can examine $100 - R/C = 9999A/(100A + B)$ and this is ≥ 0 with equality at $A = 0$. Again there are 99 solutions, which have a ratio of 100, starting with $B = 1, C = £00.01, R = £01.00$ and ending with $B = 99$, $C = £00.99$, $R = £99.00$.

Chapter 5

64. BYZANTINE SALESMANSHIP

There is a slight trick in the problem, which earlier authors often did not make clear. The common price in Venice is not the same as the common price at Padua. Let us suppose that the price in Padua was the higher — the other case simply reverses the numbers. The youngest son must have sold more pearls at the higher price in Padua than the second youngest, who must have sold more at Padua than the third youngest, ... The most that the youngest can have sold at Padua is 9 and this means that the most that the eldest can have sold at Padua is 1. But he sold at least 1 at Padua, so we see that the youngest must have sold 9 in Padua and 1 in Venice, the second youngest must have sold 8 at Padua and 12 in Venice, ... Then the value of 1 sold at Padua must be the same as the value of 11 sold at Venice. For convenience, we can take the Padua price as 11 and the Venetian price as 1, and then each son makes precisely 100.

[This problem has essentially a unique solution, but if we consider other versions of the problem, there can be many solutions. Try three sons with 50, 30, 10 pearls.]

[I believe I was the first to produce a complete theory of these problems. This appeared in *Articles in Tribute to Martin Gardner*; ed. by Scott Kim for the opening of the exhibition: Puzzles: Beyond the Borders of the Mind at the Atlanta International Museum of Art and Design, 16 Jan 1993, pp. 343–356 AND in *The Mathemagician and Pied Piper: A Collection in Tribute to Martin Gardner*; ed. by Elwyn R. Berlekamp & Tom Rodgers; A. K. Peters, Natick, Massachusetts, 1999, pp. 219–235. Although this problem is quite old, there seems to be no common name for it. The problem is treated in one form by Mahavira in c850, Sridhara in c900, and Bhaskara in 1150, but that form is less clearly expressed and has infinitely many solutions. A later form appears in Fibonacci. A version appears in the first printed English riddle collection — *The Demaundes Joyous*, printed by Wynken de Worde in 1511. Fibonacci is one of the few to give more than a single solution. He gives a fairly general rule for generating solutions, and he gives 5 solutions for the 50, 30, 10 version — but there are 55, of which 36 are positive.]

65. TESTING TIMES

Let there be a exams of length $b + \frac{1}{2}$ hours. Then we are looking for $a(b + \frac{1}{2}) = (10a + b)/2$. After some simplification, this leads to $(2a - 1)(2b - 9) = 9$. The only solutions are: $a, b = 1,9$; $2,6$; $5,5$ and $0,0$, where the last is a degenerate solution.

66. GREEK PUZZLE BOXES

If our rectangle is a by b, the first part asks for $ab = 2a + 2b$. This can be rearranged to $(a - 2)(b - 2) = 4$ and the only solutions are $a, b = 2,2$ and $3,6$. (The solution $6,3$ is just the same rectangle as $3,6$.)

For a box which is a by b by c, the volume is abc and the surface area is $2ab + 2bc + 2ac$. So we want $abc = 2ab + 2bc + 2ac$. We can and do choose a, b, c so that $a \leq b \leq c$. If $a > 6$, then also b and c are > 6, and $abc/3$ is $> 2ab$, $2bc$ and $2ac$, so the volume is greater than the surface area. Since $abc > 2bc$, we have $a > 2$. Hence a must be 3, 4, 5 or 6. Trying each value in turn, we get a quadratic equation relating b and c.

This can be factored, as in the rectangular problem, and we find ten solutions: 3,7,42; 3,8,24; 3,9,18; 3,10,15; 3,12,12; 4,5,20; 4,6,12; 4,8,8; 5,5,10; 6,6,6.

67. A MIDDLE EASTERN MUDDLE

Uncle Omar solves the muddle by loaning the estate his oil well. There are then 42 wells. The first son gets 21, the second gets 14 and the third gets 6, leaving one oil well left over which they gratefully return to Uncle Omar.

The key to the problem is that $1/2 + 1/3 + 1/7 = 41/42$, i.e. the fractions do not add up to 1. So we want integers $a \leq b \leq c$ such that $1/a + 1/b + 1/c = (d - 1)/d$, for some integer d, with a, b, c all dividing d. We rewrite this as (*) $1/a + 1/b + 1/c + 1/d = 1$. Since d is a multiple of a, b, c, we have $a \le b \le c \le d$. The solutions of this can be found by trial and error. This can be speeded up a bit by noticing that if

$$1 - 1/a - 1/b = 1/n,$$

then we want $1/c + 1/d = 1/n$. This gives

$$nc + nd = cd \quad \text{or} \quad cd - cn - dn + n^2 = n^2 \quad \text{or} \quad (c-n)(d-n) = n^2.$$

Thus these solutions correspond to the divisors of n^2.

There are 12 solution quadruples a,b,c,d: 2,3,7,42; 2,3,8,24; 2,3,9,18; 2,3,12,12; 2,4,5,20; 2,4,6,12; 2,4,8,8; 2,5,5,10; 2,6,6,6; 3,3,4,12; 3,3,6,6; 4,4,4,4. In seven of the solutions, the sons get different amounts. There are also two pseudo-solutions of (*) where d is not a multiple of a, b, c, namely: 2,3,10,15 and 3,4,4,6. If our sheik had left 1/2, 1/3, 1/10 of his wells to his three sons, then we have $1/2 + 1/3 + 1/10 = 28/30$ or 14/15, but Uncle Omar's trick would not work with an estate of 14 oil wells because 10 does not divide into 15. But the trick would work if the estate had 28 oil wells and Uncle Omar had two wells to loan.

For four sons, there are 97 solutions plus 50 pseudo-solutions.

Up until the Middle Ages, the sheik's will would have been phrased or interpreted as dividing the estate in the proportion 1/2 : 1/3 : 1/7, i.e. 21 : 14 : 6 and there would have been no need to borrow from Uncle Omar. E.g. Problem 63 of the *Rhind Papyrus* (c1700 BC) says to divide 700 loaves among four men in the proportion 2/3 : 1/2 : 1/3 : 1/4.

68. MORE PUZZLE BOXES

As before, suppose our box is $a \times b \times c$ and $a \le b \le c$. The volume of the box is abc, the surface area is $2(ab + bc + ac)$ and the sum of the edges is $4(a + b + c)$.

The first part of the problem asks for

$$(1) \quad abc = 4(a + b + c).$$

We have $4(a + b + c) \le 4(c + c + c) = 12c$, so that if $ab > 12$, then $4(a+b+c) \le 12c < abc$ and there is no solution. Hence we only consider $ab \le 12$. Also $abc = 4(a + b + c) > 4c$, so that $ab > 4$ must hold. This gives less than a dozen possibilities for a and b. For each a, b, Equation (1) gives us $c = 4(a + b)/(ab - 4)$. We try each possible pair a, b and see if c is an integer. This yields five solutions:

$$1, 5, 24; \quad 1, 6, 14; \quad 1, 8, 9; \quad 2, 3, 10; \quad 2, 4, 6.$$

The second part of the problem asks for

$$(2)\quad 2(ab + bc + ac) = 4(a + b + c),$$

or $(ab + bc + ac) = 2(a + b + c)$. Now $2(a + b + c) \leq 6c$, so that if $a + b \geq 6$, then $2(a + b + c) \leq 6c \leq ac + bc < ab + bc + ac$ and there is no solution. Hence we only consider $a + b < 6$. This already gives very few cases. Solving (2) for c gives us $c = (2a + 2b - ab)/(a + b - 2)$. There are just two solutions: 1, 2, 4 and 2, 2, 2. If $a = 0$, then there are two further solutions: 0, 3, 6 and 0, 4, 4, which give the solutions of the earlier problem for rectangles.

Actually Jessica got no solutions for either case, but that is closer to 2 than to 5, so she felt she was more successful in the second case!

69. PISTOLES AND GUINEAS

Letting G be the number of guineas and P be the number of pistoles, we want $21G + 17P = 2000$. Basic number theory gives a systematic procedure for finding such solutions, known as the Euclidean Algorithm, but here we will proceed by trial and error. We compute $2000 - 21G$ for $G = 0, 1, \ldots$. and divide the value by 17 to see if this is integral. We soon find $2000 - 7 \times 21 = 1853 = 17 \times 109$. So $21 \times 7 + 17 \times 109 = 2000$ is a solution. But there are more solutions as we can add and subtract multiples of 17×21 to the terms of this equation, e.g. $21 \times 24 + 17 \times 88 = 2000$ and we can repeat this, getting $21(7 + 17k) + 17(109 - 21k) = 2000$ for $k = 0, 1, \ldots, 5$, with the last case being $21 \times 92 + 17 \times 4$. Which case uses the fewest coins?

70. ALMOST PYTHAGOREAN

Rewrite the equation as $x^2 \pm 1 = z^2 - y^2$. The right-hand side factors as $(z - y)(z + y)$. If we know $z - y = a$ and $z + y = b$, then

$$2z = a + b, \quad 2y = b - a$$

and we see that a and b must be of the same parity, i.e. must be both even or both odd. So we can find all almost Pythagorean triangles as follows. Choose an x, form $x^2 + 1$ and factor this into two factors of the same parity. Then we do the same with $x^2 - 1$. For example, if $x = 5$, we consider $x^2 - 1 = 24 = 2 \times 12$, giving $y = 5, z = 7$, which is our student's initial triangle. The first few triangles with distinct sides found by this process are: 4,7,8; 4,8,9; 6,17,18; 6,18,19; 7,11,13; 8,31,32; 8,9,12; 8,32,33; 9,19,21; 10,49,50; 10,15,18; 10,50,51.

[My thanks to Heather Vaughan for mentioning her student's problem, which led to the above problem. We published it in *Mathematics Magazine*. 60:4 (1987) 244 & 248.]

71. ALMOST PYTHAGOREAN TRIANGLES ON A GEOBOARD®

Sadly, there aren't any such triangles on a Geoboard®. Consider any triangle on the Geoboard®. Take any vertex as the origin of coordinates. Then the other two vertices are at positions (a, b), (c, d), where a, b, c, d are all integers. If we let the lengths of the sides be x, y, z, then their squares are given by $x^2 = a^2 + b^2$, $y^2 = c^2 + d^2$ and $z^2 = (a - c)^2 + (b - d)^2$. Then $x^2 + y^2 - z^2 = 2(ac + bd)$, hence is an even integer, and so cannot be ± 1. This holds no matter which vertex is taken as the origin so we do not get an almost Pythagorean triangle, no matter which side is taken as the 'almost hypotenuse'.

72. GEOBOARD® TRIANGLES AGAIN

When we attach coordinates to the Geoboard® points, the coordinates are integers. The distance between two points is given by the Theorem of Pythagoras, so the distance squared is the sum of two squares of integers. If a triangle is in two really different positions, then some side has to occur in two really different positions, i.e. the square of its length must be a sum of two squares in two really different ways. That is, we do not consider $5 = 1^2 + 2^2 = 2^2 + 1^2$ as two different representations. Making a table of sums of two squares, we find that the first few numbers which are sums of two squares in two different ways are the following.

$$\begin{aligned}
25 &= 5^2 + 0^2 = 4^2 + 3^2 \\
50 &= 7^2 + 1^2 = 5^2 + 5^2 \\
65 &= 8^2 + 1^2 = 7^2 + 4^2 \\
85 &= 9^2 + 2^2 = 7^2 + 6^2 \\
100 &= 10^2 + 0^2 = 8^2 + 6^2
\end{aligned}$$

Trying these lengths in various ways leads to two simple examples.

Vertices at (0, 0), (5, 0), (0, 5) and vertices at (0, 0), (4, 3), (−3, 4) both form an isosceles right triangle of sides 5, 5 and $\sqrt{50}$.

Vertices at (0, 0), (5, 0), (6, 7) and vertices at (0, 0), (8, −1), (3, 4) both form a triangle of sides 5, $\sqrt{50}$, $\sqrt{65}$.

Sometime after finding the above, I was surprised to find that there are even simpler examples.

Vertices at (0, 0), (5, 0), (4, 2) and vertices at (0, 0), (4, 2), (3, 4) both give a right triangle of sides $\sqrt{5}$, $\sqrt{20}$, 5. I found this by taking the midpoint between (5, 0) and (3, 4). The midpoint of (7, 1) and (5, 5) gives another example with sides $\sqrt{5}$, $\sqrt{45}$, $\sqrt{50}$. Some computer assistance turned up another type of example.

Vertices at (0, 0), (5, 0), (7, 1) and at (0, 0), (4, 3), (5, 5) both give triangles with sides $\sqrt{5}$, 5, $\sqrt{50}$.

[My thanks to Tom Henley for asking how many distinct triangles there are on a Geoboard®. Unfortunately, due to the phenomenon described in this problem, we are unable to solve the original question.]

73. THE CCCC RANCH

Yes, she can. Let a be the number of copperheads, b be the number of chickens, c be the number of children and d be the number of cows. Then we know $a+b+c+d = 17, a+b+d = 11, 2b+2c+4d = 50$. Clearly, we have $c = 6$, so the equations reduce to $a+b+d = 11, 2b+4d = 38$. The latter gives $b+2d = 19$, which has a number of solutions: $b,d = 19,0$; $17,1$; …; $3,8$; $1,9$. The first equation requires $b+d < 11$ in order for a to be greater than 0. The only solution is then $b, d = 9, 1$, giving $a = 1$.

[This is an example of The Hundred Fowls Problem, which first appears in the *Suan Ching* of Chang Chiu-Chien in about 475. There a man went to market and bought 100 fowls for 100 coins. Roosters cost 5, hens cost 3 and chicks cost 1/3. (Such problems also appear in the Indian *Bakhshali Manuscript* which has been variously dated between the 2nd and 12th centuries, and these problems have been common exercises ever since.) Ambitious readers may like to find all the solutions of Abu Kamil's problem of c900. A man goes to market and buys 100 fowls for 100 coins. Ducks cost 2, doves cost 1/2, ring-doves cost 1/3, larks cost 1/4 and sparrows cost 1.]

74. DOUBLING UP

It is easiest to work backwards from the final state at which they each have M matchsticks. Note that the poorer one must have won each hand, otherwise play would have ended earlier on. On the last hand, the poorer one (Sophie) must have won as much as she had. So Sophie must have had $M/2$ and Jessica had $3M/2$. That is, the poorer one has doubled her money and previously had half as much. In the next to last hand, Jessica must have won $3M/4$ and she must have had $3M/4$ to Sophie's $5M/4$. Carrying on, we get the following fractions occurring, where I have omitted the common factor of M.

Jessica	1	3/2	3/4	11/8	11/16	43/32	43/64
Sophie	1	1/2	5/4	5/8	21/16	21/32	85/64

The last fractions multiplied by M must be whole numbers, so the least value of M is 64 and they would have had 43 and 85 matches to start with.

[To find the general solution, I found it easiest to consider the sequence of larger values, after backing up n hands, $n = 0, 1, \ldots$:

$$1, 3/2, 5/4, 11/8, 21/16, 43/32, 85/64, \ldots, a_n/2^n, \ldots$$

and the sequence of smaller values:

$$1, 1/2, 3/4, 5/8, 11/16, 21/32, 43/64, \ldots, b_n/2^n, \ldots.$$

One quickly sees various properties of the sequences of numerators. We have $b_n = a_{n-1}$ because the richer loses half his money and becomes the poorer at each stage. Also $a_n = 2b_{n-1} + a_{n-1}$ because the poorer gains half of what the richer had to become the richer. From these, we obtain the following relations:

$$a_n = a_{n-1} + 2a_{n-2}; \quad a_n = a_{n-2} + 2^n; \quad a_n + a_{n-1} = 2^{n+1}.$$

These allow us to compute a_n and b_n easily. From the theory of recurrence relations, one can solve this explicitly as

$$a_n = [2^{n+2} - (-1)^{n+2}]/3$$

and this is the just the integer nearest to $2^{n+2}/3$. It is slightly easier to consider the sequence $a_{n-2.}$ for $n = 0, 1, 2, \ldots$, which is: $0, 1, 1, 3, 5, 11, 21, 43, 85, \ldots$].

75. A STRANGE CHESSBOARD

An N by N chessboard has $4(N-1)$ edge squares and $(N-2)^2$ interior squares. Equating these gives us $N^2 - 8N + 8 = 0$. The solution of this is $N = 4 \pm 2\sqrt{2}$. The minus sign gives a negative value, which makes no sense in the problem. The plus sign gives us $N = 6.828\ldots$, which is a rather awkward size of board, though this is the answer given by Maestro Biagio. However, we can reinterpret the problem as asking for a square area (e.g. a garden) such that a strip of width one around its edge covers exactly half the area.

For a chessboard of side N, if we take $N = 7$, then there are 24 edge squares and 25 interior squares, which is as close as you can get with a square board. But did you think to try a non-square board??

An M by N board has $2M + 2N - 4$ edge squares and $(M-2)(N-2)$ interior squares. Equating these gives us $MN - 4M - 4N + 8 = 0$. Adding 8 to each side allows us to factor, getting $(M-4)(N-4) = 8$. The only whole number solutions are $M, N = 5, 12;\ 6, 8$ and the reversals of these.

[I later found that the M by N problem appears in: Andy Bernoff & Richard Pennington; Problems Drive 1984; *Eureka* 45 (Jan 1985) 22–23 & 50, Prob. 5.]

76. A POCKET MONEY PROBLEM

Let A, B, C be the amounts the girls have. We know that $A = (B+C)/3$ and $B = (A+C)/2$. One can proceed directly to express C in terms of $A+B$, but it is easier to make the equations more symmetric by adding $A/3$ and $B/2$ respectively to get $4A/3 = (A+B+C)/3$ and $3B/2 = (A+B+C)/2$. Let T denote the total $A + B + C$. Then we have $A = T/4$, $B = T/3$, so $C = T - A - B = T(1 - 1/4 - 1/3) = 5T/12$. If $C = X(A + B)$, then $(1 + X)C = XT$ and $C = XT/(1 + X)$. So we want the X such that $X/(1 + X) = 5/12$, or $12/5 = (1 + X)/X = 1 + 1/X$, so $7/5 = 1/X$ and $X = 5/7$. Thus the girls' monies are in the proportion $1/4 : 1/3 : 5/12 = 3 : 4 : 5$ and the least amounts they could have had are 3, 4, 5.

77. TAKE YOUR SEATS

Let x, y, z denote the number of seats at £3, £4, £5. Then

$$x + y + z = 2000$$

and the receipts of £7500 gives us: $3x + 4y + 5z = 7500$. Subtracting three times the first equation from the second leaves us with $y + 2z = 1500$. Since both 1500 and $2z$ are even, y must also be even, let us say $y = 2w$. Then the last equation reduces to $w + z = 750$. For each value of $w = 0, 1, 2, \ldots, 750$, there are corresponding values of $y = 2w = 0, 2, 4, \ldots, 1500$; $z = 750, 749, 748, \ldots, 0$ and $x = 1250, 1249, 1248, \ldots, 500$. Thus there are no less than 751 possible solutions, even assuming there are 2000 seats.

If the number of seats is not specified, we have only our second equation: $3x + 4y + 5z = 7500$. There is no easy way to determine all the solutions. I have a computer program that does such problems and it ran for about an hour while I did the calculations by hand, taking about ten minutes longer. We agreed that there are no less than 469,501 solutions of the original problem. So giving just one solution is a bit of an understatement!

[If you are interested, there are 468,001 solutions with only positive values.]

78. A MILLENNIUM CONUNDRUM

To solve $19X + 99Y = 2000$, one needs to find some solution X, Y. After some fiddling, I observed that $99 - 19 = 80$ and 80 divides into 2000 twenty-five times, so we have $99 \cdot 25 - 19 \cdot 25 = 2000$. Now if we subtract $99 \cdot 19$ from the first term and add the same amount, as $19 \cdot 99$ to the second term, we get $99 \cdot 6 + 19 \cdot 74 = 2000$. From the fact that 19 and 99 have no common factor, it can easily be shown that all solutions are obtained from one solution by adding and subtracting multiples of $19 \cdot 99$. However, repeating the process makes X negative, so only the one case already seen has both X and Y being positive.

79. A MAGIC CUBE

For convenience, we label all the vertices. Let A, B, C, D be the top vertices, going clockwise. Let E, F, G, H be the vertices underneath the top vertices. Then A is adjacent to B, D and E. Then $A+B+C+D = C+D+G+H$, so $A + B = G + H$, and similarly for any pair of diagonally opposite edges, as asked in the Olympiad problem. Now let us rotate the cube so that $A = 1$. Then $A + B \leq 1 + 8 = 9$. Now suppose either G or H has the value 8. The other value must then be > 1, so $G + H \geq 2 + 8 = 10$, so $A + B \neq G + H$ if G or H is 8. Symmetrically we also see that neither C nor F can be 8. Thus 8 must occur at B or D or E, i.e. at one of the vertices adjacent to $A = 1$. We can thus assume $E = 8$. Then $A + E = 9 = C + G$.

Now we need a little more information. Consider the sum of all values $A + B + \cdots + H$. These are some permutation of the integers $1, 2, \ldots, 8$, so the sum is the same as $1 + 2 + \cdots + 8 = 36$. Further, the top face, $A + B + C + D$ and the bottom face, $E + F + G + H$ both have the same 'magic sum', say S. But then $2S = A + B + \cdots + H = 36$, so that our magic sum is fixed as 18. Now consider the front face. This gives us $A + D + E + H = 18$ and we know $A + E = 9$, so $D + H = 9$. Similarly, or by use of the Olympiad result, we have $B + F = 9$. Thus each vertical edge of our cube has its vertex values adding to 9. This automatically makes all side faces add to the magic sum of 18. But there are only 4 ways we can make 9 with two of our integers, namely:

$$1 + 8, \quad 2 + 7, \quad 3 + 6, \quad 4 + 5,$$

so our vertical edges must consist of these pairs. We must also have $A + B + C + D = 18$. This requires us to take one value from each of our pairs in such a way as to total 18. Starting with $A = 1$, we see that we cannot choose 2 to be in the upper face, as then the maximum value will be $1+2+6+5 = 14$. So 7 must be in the upper face and the other two values must add up to 10, which requires 6 and 4 to be in the upper face. Thus a solution has 1, 4, 6, 7 in the upper face with all vertical edges totaling 9. Any arrangement of 1, 4, 6, 7, in the upper face works. If we fix $A = 1$, as we have done, then there are 6 arrangements of 4, 6, 7. Half of these are equivalent to the other half under reflection, so one might say there are three distinct solutions, determined by which of 4, 6, 7 is opposite to A in the upper face.

Chapter 6

80. SKELETON DIVISION IN THE CUPBOARD

The solution of the problem is 16 divided into 1062 yielding 66.375 and this is unique.

Let A be the divisor and B be the dividend. Since B/A has exactly three decimal places, then B/A is an integral multiple of 1/1000. However, B/A may not be in lowest terms, so let $B'/A' = B/A$ but with B' and A' having no common factor. Then A' actually divides 1000. The denominators A' which give exactly three decimals are then: 8, 40, 200, 250, 500, 1000. But A' has at most two digits — because A, which is a multiple of A', has two digits. Hence $A' = 8$ or 40 and so $A = 8C$ where $C = 2, 3, \ldots$, or 12.

Now $1/8 = .125$ and $1/40 = .025$ and any multiple of these which has three decimals will have the last decimal digit being a 5. But the last stage of the division then shows us that $5A$ has only two digits, i.e. $5A < 100$, so $A < 20$, and we must have $A = 16$ and so $A' = 8$. Since $B/16 = B'/8$, where B' must be odd, we have that B is twice an odd number.

Now the skeleton of our division is as follows.

```
        ef.gh5
   16 ) abcd
        jk
        ---
        lmd
         pq
         ---
          r.0
          s.t
          ---
          u.v0
          w.xy
          ----
             80
             80
             --
```

Since $abc \geq 100$, but $jk < 100$, then jk must be the largest two-digit multiple of 16, namely $6 \times 16 = 96$, and $abc < 7 \times 16 = 112$. Since $lm \geq 10$, we have $abc \geq 106$. Similarly, we see that $pq = 96$ and since $r < 10$, we have $lmd < 106$. Hence we must have $lm = 10$ and $abc = 106$ and $abcd < 1066$. Now the only number which is twice an odd number in the range 1060 to 1065 is 1062, and this is seen to completely fit the given skeleton.

81. A PAN-DIGITAL FRACTION

The unique solution is $5/34 + 7/68 + 9/12 = 1$.

82. EVEN MORE MIXUPS

Recalling that a two-digit number ab has the value $10a + b$, the problem $ab - cd = ba/cd$ gives us $[(10a + b) - (10c + d)](10c + d) = 10b + a$. Multiplying this out leads to no cancellations, but examination of the units digits shows us that $(b - d)d = a$, up to adding multiples of 10. [Mathematicians say $(b - d)d$ is congruent to a, modulo 10, and write this as $(b - d)d \equiv a \pmod{10}$.] This allows us to do a systematic search — take a value of a and add some multiple of 10 to it to get $a' = a + 10k$. Factor a' as $d(b - d)$ and use this to get b. Putting a, b, d into the problem allows us to easily determine the possible values for c.

At first, it looks like we have to treat 100 values of a'. However

$$d + b - d = b$$

is a single digit, hence is at most 9. The relation $xy \leq [(x + y)/2]^2$ is easily seen by multiplying out. [Mathematicians know this as the arithmetic-geometric mean inequality.] Applying with $x = d$, $y = b - d$, we have $(x + y)/2 = b/2 \leq 4\,1/2$, hence $a' = bd \leq 20\,1/4$, so a' is at most 20. In many cases, the factors of a' are few and are often too big to use in our problem, so it only takes some minutes to check all 30-some cases and find there are just four solutions:

$$27 - 3 = 72/3; \quad 27 - 24 = 72/24; \quad 49 - 2 = 94/2; \quad 49 - 47 = 94/47.$$

Rewriting the first as $27 = 3 + 72/3 = 3 + 24$, we see that we have factored 72 into a product of factors which add up to 72 reversed. The first two of our solutions correspond to the same factorization, taken in opposite orders, and the last two are similarly related.

83. TWO WRONGS CAN MAKE A RIGHT

Solutions are given by the following 21 values of *WRONG*, in ascending order, though it's unlikely you'll find them in this order.
12734, 12867, 12938, 24153, 24765, 25173, 25193, 25418, 25438, 25469, 25734, 25867, 25938, 37081, 37091, 37806, 37846, 37908, 49153, 49265, 49306.

I found 14 solutions by hand, but one had a repeated value and then I found I had overlooked one in copying. The others were simply missed by failing to check all cases carefully. I think the original author was a bit lazy in not looking for more solutions, but I can now understand why he didn't try to find all solutions.

$O = 0$ occurs for the cases: 37081, 37091. $I = 1$ occurs for 25734, 25867.

84. THREE LEFTS MAKE A RIGHT

I found one solution by hand and saw that I had only gone through a small part of the possible cases, so I wrote a small program to find all cases. *LEFT* can have the 27 values: 8910, 5820, 5430, 8730, 5830, 5940, 9160, 9180, 6580, 8190, 8390, 8305, 7805, 8905, 6915, 8915, 6325, 3625, 4625, 9135, 6235, 9045, 6245, 6945, 8065, 7695, 6795.

There are just 4 values where the digit 0 does not occur: 8915, 6325, 4625, 6245. In no case does $I = 1$.

85. OXO CUBE

The answers are: $178^3 = 5639752$ and $189^3 = 6751269$.

If *NOTHING* × *NOTHING* is a cube, then *NOTHING* must be a cube. The integers whose cubes have seven digits are: 100, 101, ... , 215. So I asked my computer to compute all these cubes and examine their digits. In 13 cases, the two *N* digits were the same, but only in the cases given above were the other digits all distinct. One could scan a table of cubes to find the answers. Starting with 100^3, the cubes have first digit 1 until 125, so it is easy to scan the tens digit to see if it is a 1 and then check the other digits.

86. STIR TWO WHEAT

In the most normal case (no repetitions and no leading zeroes), the left end of the sum $STIR + TWO = WHEAT$ shows that $W = 1$, $S = 9$, $H = 0$ and there must be a carry from $T + T$. Now $I + W = I + 1$. This can exceed 9 only if $I = 9$, which is a repeat, or if $I = 8$ and there is a carry from $R + O$, but this leads to $A = 0$, which is also a repeat. Hence there cannot be a carry to $T + T$, but there is a carry from it, so $T = 6, 7, 8$. We examine each case.

$T = 6$, $E = 2$. Then R, O (in either order) must be one of the pairs 0, 6; 1, 5; 2, 4; 7, 9, but each of these repeats some value.

$T = 7$, $E = 4$. Then R, O (in either order) must be one of the pairs 0, 7; 1, 6; 2, 5; 3, 4; 8, 9, but only 2, 5 avoids a repetition. Then $A = I + 1$ and so we need I, A to be two consecutive unused digits, but only 3, 6, 8 are unused.

$T = 8$, $E = 6$. Then R, O (in either order) must be one of the pairs 0, 8; 1, 7; 2, 6; 3, 5, but only 3, 5 avoids repetition and again there are no two consecutive unused digits for I, A.

So there are no solutions in the normal case. If we let $W = 0$, we see that $H = S + 1$ and there must be a carry from $T + T$. Also, there must be a carry from $R + O$ and $A = I + 1$. Since $A \neq 0$, we cannot have a carry to $T + T$. Since $H \neq 0$, $T > 5$. Since T arises from $R + O$, and the largest sum of two distinct digits is 17, we have $T \leq 7$, so $T = 6$ or 7.

$T = 6$, $E = 2$. Then $R + O = 16$ and R, O (in either order) must be 7, 9, leaving 1, 3, 4, 5, 8 unused. But we need both S, H and I, A as consecutive unused digits and this cannot be done here.

$T = 7$, $E = 4$. Then R, O (in either order) must be 8, 9, leaving 1, 2, 3, 5, 6 unused. So the pairs S, H and I, A can be 1, 2 and 5, 6 or 2, 3 and 5, 6 or the order of the pairs can be reversed. Since R, O can be either way round, we get 8 solutions, where $STIR = 1758, 1759, 2758, 2759, 5718, 5719, 5728, 5729$.

In the general case where repetitions and leading zeroes are permitted, we can let R, O, I, S be any digits and the rest of the values are determined, so there are 10,000 solutions! If we prohibit leading zeroes, then $S = 9$ is determined and we must have $T \geq 5$, which means that $R + O$ can take on just half of its values, i.e. just 50 pairs can occur. Then I can take on 10 values and we have 500 solutions in this case — rather more than the single solution given in the book! Finally, if only the digits 0, 1, 9 occur in this case, then $T = 9$, so $R, O = 0, 9$ or $9, 0$. Since $I + W = I + 1$ gives a carry to $T + T$, we must have $I = 9$, $A = 0$ and there are just two solutions: $9990 + 919 = 10909$ and $9999 + 910 = 10909$.

87. FLY FOR YOUR LIFE

Assuming $O = 0$ and $I = 1$, let $C2$ be the carry from $L + 0 + U$. This can be at most 2. Then $C2 + F + F + 0 = 1$ or 11 or 21. This rules out $C2 = 0$ or 2. Also $F \neq 0$, so this sum cannot be 1 and the sum is at most 19, so we deduce that this sum must be 11 and hence $F = 5$, $C2 = 1$ and $L = Y + 1$ (so $Y \leq 8$). Now let $C1$ be the carry from $Y + R + R$, which can be 0, 1 or 2. We have $C1 + L + U = 15$ or $C1 + Y + U = 14$. For each value of Y and $C1$, this determines U. But $C1 \leq 2$ and $U \leq 9$ imply that $Y \geq 3$.

For $Y = 3$, we must have $C1 = 2$, i.e. $Y + R + R \geq 20$, which can only occur for $R = 9$.

For $Y = 4$, we must have $C1 = 1$ or 2, which implies $R = 3, 4, 5, 6, 7, 8, 9$.

For $Y = 5, 6, 7, 8$, we can have any R.

This gives us 48 solutions without leading zeroes. The only case with distinct letters having distinct values is: $598 + 507 + 8047 = 9152$.

[If $F = 0$, then $L = Y$ and $C1 + Y + U = 10$.

$Y = 0$ requires $C1 = 1$ and $R = 5, 6, 7, 8, 9$.

$Y = 9$ requires $C1 = 0$ or 1 and $R = 0, 1, 2, 3, 4, 5$.

$Y = 1, 2, 3, 4, 5, 6, 7, 8$ allows any R.

So there are 91 solutions with leading zeroes and a total of 139 with repetitions allowed.]

88. TWO ODDS MAKE AN EVEN

$ODD + ODD = EVEN$. Since $EVEN \leq 988 + 988 < 2000$ and $E \neq 0$, it follows that $E = 1$. Further, since $E \neq N$, it follows that there must be a carry from the units column to the tens column, so $D \geq 5$ and $E = N + 1$. But since $E = 1$, we have $N = 0$, so $D = 5$. This leaves us with $1 + O + O = 1V$, so $O > 5$ and we can pick any such O which gives a value of V distinct from the other letters. There are just two cases: $O = 6$ and $O = 8$, giving $655 + 655 = 1310$ and $855 + 855 = 1710$.

$ODD + ODD + ODD = EVEN$. Since $EVEN < 3000$ and $E \neq 0$, it follows that $E = 1$ or 2. Again, there must be a carry from the units column, so $D \geq 4$. I can't see any simpler argument here than to inspect the triples of the possible values of DD. Three times 44, 55, 66, 77, 88, 99 is 132, 165, 198, 231, 264, 297. In no case is the tens digit either 1 or 2, so there are no solutions here.

89. ODDS AND EVENS — 2

$4 * ODD = EVEN$. Since $EVEN < 4000$ and $E \neq 0$, it follows that $E = 1$ or 2 or 3. Since there must be a carry from the units column, we have $D \geq 3$. The last two digits of four times 33, 44, 55, 66, 77, 88, 99 are 32, 76, 20, 64, 08, 52, 96. So D must be either 3 or 5, with corresponding E of 3 or 2. But $D = 3$ is making $E = 3$, which is not permitted, so we need only consider $D = 5$, $E = 2$. There is only one unused value of O which makes $4*ODD$ have a value between 2000 and 2999, namely 6. But we have $4*655 = 2620$, which has a repeated value, so there are no solutions here.

$5*ODD = EVEN$. As before, we see that $E = 1, 2, 3, 4$ and $D \geq 2$. The last two digits of five times 22, 33, ..., 99 are 10, 65, 20, 75, 30, 85, 40, 95. So only $D = 2, 4, 6, 8$ can occur and in all cases $N = 0$. In each case, there is only one unused value of O which makes $5*ODD$ have a value between $E000$ and $E999$, namely $O = D + 1$. The last case has a repeated value, leaving three solutions: $ODD = 322, \;\; 544, \;\; 766$.

90. ODDS AND EVENS — 3

$6*ODD = EVEN$ has the unique solution $ODD = 655$.
$7*ODD = EVEN$ has two solutions: $ODD = 722$ and 822.
$8*ODD = EVEN$ has four solutions: $ODD = 922, \;\; 644, \;\; 744, \;\; 366$.
$9*ODD = EVEN$ has no solutions.

There are 32 different values of the multiplier which give solutions, ranging from $k = 2$ to $k = 77$. There are 52 solutions. Twenty multipliers have unique solutions; six multipliers have two solutions; four multipliers have three solutions; two multipliers (8 and 24) have four solutions.

There are 7 solutions of $K*ODD = EVEN$ and 9 solutions of $KL*ODD = EVEN$. The smallest case with a two-digit multiplier is

13*544 = 7072. There are also three solutions of $KK^*ODD = EVEN$: 33*199 = 6567; 44*133 = 5852; 77*122 = 9394.

91. THREE SQUARES AGAIN

For any given solution, a hundred times it is also a solution. For example, from 4, 9, 49, we get 4, 900, 4900. As I said, these are easily found and can be described as having trailing zeroes.

The eight solutions up to 10,000, eliminating those with leading and trailing zeroes, are the following: 4, 9, 49; 16, 9, 169; 36, 1, 361; 1, 225, 1225; 144, 4, 1444; 16, 81, 1681; 324, 9, 3249; 4, 225, 4225. One has just to look at the squares of numbers from 1 to 100 and it's pretty easy to recognize when such a square splits into two squares, though it helps to have a table of squares to refer to.

My computer finds: 0 solutions up to 10; 1 solution up to 100; 3 solutions up to 1000; 8 solutions up to 10,000; 14 solutions up to 100,000; 20 solutions up to 1,000,000; 36 solutions up to 10,000,000; 46 solutions up to 100,000,000; 66 solutions up to 1,000,000,000; 88 solutions up to 10,000,000,000; 125 solutions up to 100,000,000,000; and 147 solutions up to 1,000,000,000,000. I have not been able to see any patterns in these solutions.

The most extraordinary example found is the pair: 4, 950625, 4950625; 49, 50625, 4950625, where two solutions have the same concatenated value!

92. A FOOTBALL GAME PROBLEM

First we see that the hundreds column gives us $O + A = A$. My initial reaction was that this can only happen if $O = 0$ and there is no carry from the tens column. I proceeded to solve the problem under this assumption. However, Toni Beardon, the then Director of the NRICH club (http://nrich.maths.org) told me she recalled there being more solutions. I then realized that one can have $O = 9$ if there is a carry from the tens column. This gives us two cases and we analyze each of them.

Case 0: $O = 0$. From the tens column, we have $0 + L = M$ and this can only happen if there is a carry from the units column and $M = L + 1$. This tells us that $T + L \geq 10$, but since 0 is already used, we must have $T + L \geq 11$, which requires T to be at least 2. However, we also have $M \leq 9$, so $L \leq 8$ and so $T \geq 3$.

Looking at the rest of the problem, we see it is basically independent of what we have just done, except that the letters F, B, G, A must be positive and distinct from the letters T, L, E, M, and that, from the thousands column, we have $F + B = G \leq 9$.

Case 9: $O = 9$. The tens column gives us $C + 9 + L = 10 + M$ where C is the carry from the units column. $C = 1$ makes $L = M$, so we know $C = 0$ and $L = M + 1$. This only gives us $1 \leq L \leq 8$. The units column tells us that $T \geq 1$ (else $L = E$) and $C = 0$ also tells us that $L + T \leq 9$, so that $T \leq 8$.

Looking at the rest of the problem as before, we see that the letters F, B, G, A must be positive and distinct from the letters T, L, E, M, and that $1 + F + B = G \leq 9$.

A simple program to find solutions will have to loop through the possible values of T, L (which determine E and M), then through the possible values of A, F, B (which determine G). This can be done separately for $O = 0$ and $O = 9$ or one can combine the two by another loop — to be on the safe side, I wrote a program which looped through all values of O! In no case do the constraints on T and L make a huge difference to a program, nor did the extra loop. There are 172 solutions with $O = 0$ and 52 solutions with $O = 9$.

[My mistake points out a common problem — one leaps to a initial result and then gets so tied up with the solution that one doesn't check the initial result again!]

93. SOME PAN-DIGITAL SUMS OF FRACTIONS

The minimum value occurs at

$1/74 + 2/85 + 3/96 = 0.06829\ 29252\ 78219 \ldots$.

The maximum value occurs at

$7/46 + 8/25 + 9/13 = 1.16448\ 16053\ 51171 \ldots$.

Observe the almost regular distributions of the digits. I was surprised that the maximum value is only a little greater than one.

We can deduce most of the solutions. We assume $A < D < G$ in order to avoid permutations of the summands. Now consider the minimum sum. If we have a fraction A/BC occurring, then it must have $A < C < B$, otherwise we could rearrange the three digits to get a smaller value. Hence 1 must be a numerator and so $A = 1$. If 2 is in a denominator, then exchanging it with

D will decrease the sum, so we must have $D = 2$ and, similarly, $G = 3$. We now must have $BC < EF < HI$, which combines with $C < B$, $F < E$, $I < H$ to show $H = 9$. There are now only a few cases left to consider though they are not quite so simple to deal with, but careful checking reveals that the given solution is the minimum.

Similar arguments work for the maximum, giving $B < C < A$, etc., leading to $G = 9$, $D = 8$, $A = 7$ and then $HI < EF < BC$ leads to $H = 1$, but the ultimate solution is not quite as regular as for the minimum.

94. MORE PAN-DIGITAL FRACTIONS

Since I still had the program for the previous version of the problem, it was easy to modify it for this version and find the unique solution is

$$1/(3 * 6) + 5/(8 * 9) + 7/(2 * 4) = 1.$$

I don't know if anyone has managed to solve this by hand?

95. EVEN MORE PAN-DIGITAL FRACTIONS

Since I still had the program for the previous version of the problem, it was easy to modify it for this version and find there are two solutions

$$2/(6 + 9) + 3/(7 + 8) + 4/(1 + 5) = 1$$

and

$$2/(7 + 8) + 3/(6 + 9) + 4/(1 + 5) = 1.$$

Note that when we do the sums, the first two denominators are the the same, which is why I said the solution was essentially unique.

Has anyone managed to solve this by hand?

96. SEVEN DAYS MAKE ONE WEEK

My computer finds nine solutions. Only two of these have $D \neq 0$, which are the solutions given by the previous author. The values of *DAYS WEEK*, in ascending order are: 0254 1778, 0269 1883, 0381 2667, 0461 3227, 0524 3668, 0762 5334, 0921 6447, 1048 7336, 1207 8449. Note that the value of E does not uniquely determine the solution unless $D \neq 0$ is also assumed.

Chapter 7

97. A SNEAKY SEQUENCE

The sequence is the representation of 16 in base b, for $b = 16, 15, \ldots, 3$. Hence the next term is 16 in base 2, namely 10000, and the following term is 16 in base 1.

Base 1 is not a common base because it doesn't have any obvious meaning. We would expect to write 16 as a sum of terms of the form $a_i 1^i$ with $0 \leq a_i < 1$. This only allows a_i to be 0, which makes all terms of our sum equal to 0, which is clearly not very useful. If we adjust the range on a_i to be $0 < a_i \leq 1$, which only allows a_i to be 1, then we want to represent 16 as a sum of terms which are all 1 and we can write $16 = 1111111111111111$ (with 16 ones) as a kind of base 1 representation. That is, base 1 can be viewed as the primitive technique of tallying.

98. WHATEVER NEXT?

A. 1000, if you pronounce this 'a thousand'. But if you pronounce this 'one thousand', then the answer is 2000. This is the sequence of numbers which have no letter 'e' in their names.

B. 100. This is the sequence of numbers with no letter 't' in their names.

99. SENT FOR A TEASER!

The next letter is F. Indeed the next twenty letters are Fs, followed by twenty Ss, ten Es and ten Ns. They are the first letters of the integers, starting at seventeen, just to confuse you and produce an interesting puzzle. But what comes next? That depends on how you say 100, 101, ... If you say 'one hundred, one hundred one, . . . ', then there are 100 Os. But you might say 'a hundred' or just 'hundred'.

I used this problem in *The Telegraph* on 1 Aug 1998. In October 2014, I received a letter from a desperate reader who had not seen the answer!

100. A PERVERSE SEQUENCE

Ten, Nine, Eight, Seven, Six, Five, Four, Three, Two, One could be followed by Z for Zero, but Werner probably intended B for Blast-off! F for Fire! and L for Lift-Off! are other possibilities and Americans might use G for Geronimo! I forgot to say Jessica's cousin's surname is von Braun.

101. A PENROSE SEQUENCE

If one tries the usual techniques, one gets nowhere. If you are familiar with differences, you will take the differences, but these do not lead anywhere. [If you are really familiar with differences, you may realize that not getting anywhere often indicates that there is some exponential behavior.] Most people now factor the numbers — but the only one with an interesting factor is 744 which has a prime factor of 31. This should be immediately recognized as $32 - 1 = 2^5 - 1$ and so we try to factor the terms by $2^x - 1$ and we get the following:

Exponent	$x = -3$	-2	-1	0	1	2	3	4	5	6
Term	35	45	60	P	120	180	280	450	744	1260
divided by	$-7/8$	$-3/4$	$-1/2$	0	1	3	7	15	31	63
equals	-40	-60	-120	??	120	60	40	30	24	20

and one easily sees that the quotient is $120/x$, so the desired expression is $120(2^x - 1)/x$ for $x = -3, -2, \ldots, 6$, which is readily extended — but the extended values are no longer integers!

But there is a further sting in the problem. The missing term P corresponds to $x = 0$ and the resulting expression is $120 \cdot 0/0$, which is indeterminate. To evaluate this, one must use limiting techniques from elementary calculus, of which the easiest is L'Hospital's Rule, which tells us that in this situation, $\lim_{x\to 0} 120(2^x - 1)/x$ is the same as the limit of the ratio of the derivatives of the numerator and the denominator. Recalling that the derivative of 2^x is $(\log 2)\, 2^x$, where the natural logarithm is used, we have that the desired value is $P = \lim_{x\to 0} 120(\log 2)2^x/1 = 120 \log 2 =$

$120 \times 0.693147\ldots = 83.17766\ldots$, which I suggest calling 'Penrose's Number'. This is truly an irrational problem that drives one to the limit!!

One can also look at the ratios of consecutive terms. These will converge to 2, but the limit is not obvious from the terms given.

102. MORE SEQUENCES

A. These are the final letters of the numbers, 1, 2, 3, ...

B. These are the numbers in reversed alphabetic order — i.e. in the alphabetic order of their reversals: eerht, enin, eno, evif, ... I've omitted orez as some people would spell it ho or thguon.

C. If you spell out the numbers, each entry begins with the last letter of the preceding entry. Both the ?s thus must start with e and end with t and the obvious answer is 8 for both cases, though 88 and other values could be used.

103. EVEN MORE SEQUENCES

A. These are the numbers of the months when arranged in alphabetical order.

B. These are the points where the form of compound numbers changes from having the unit first to having the unit last.
Greek: *δώδεκα – δεκατρια* (dódeka — dekatria).
Spanish: quince — diez y seis.
French: seize — dix-sept.
Italian: sedici — diciasette.
English: nineteen — twenty(?) — twenty-one.
Russian: девятнадцать — двадцать(?) — двадцать один
(devyatnadtsat' — dvadtsat' (?) — dvadtsat' odin).
In Arabic, German, Hebrew and Norwegian, though the words for 11 and 12 are a bit irregular, the basic forms are one-and-ten, . . . , nine-and-ten, twenty, one-and-twenty, . . . , though the 'and' may be elided or completely omitted. The pattern usually gets disrupted at 100. There are several languages where the unit is last from 11 on, e.g. Chinese, Hungarian, Turkish. Latin is confused because 18 and 19 are two-from-twenty and one-from-twenty as well as ten-and-eight and ten-and-nine, and 17 is usually seven-ten but is sometimes ten-and-seven.

[The Mayans changed at 13, but the other way round, from ten-two to three-ten.]

C. These are the numbers with no Ns in their names. The last number in this sequence is 88.
D. These are the numbers with no Is in their names. The last such is 777,777.
E. These are the numbers with no Os in their names. The last such is 999.
F. These are the smallest numbers with 3, 4, 5, ... letters in their names.

104. EVEN MORE SEQUENCES — 2

These are the numbers with no Ns in their names. The last number in this sequence is 88. It is not hard to form the complete list.

2, 3, 4, 5, 6, 8,
12,
30, 32, 33, 34, 35, 36, 38,
40, 42, 43, 44, 45, 46, 48,
50, 52, 53, 54, 55, 56, 58,
60, 62, 63, 64, 65, 66, 68,
80, 82, 83, 84, 85, 86, 88,

So there are just 42 of these.

105. EVEN MORE SEQUENCES — 3

These are the numbers with no Os in their names. The last such is 999. To count them, we first note that the hundreds digit will give no trouble and can be 3, 5, 6, 7, 8, 9. So we examine the numbers from 1 to 99. In 1–9, we get 3, 5, 6, 7, 8, 9, but in 10–19, we get 10, 11, 12, 13, 15, 16, 17, 18, 19. From 20 to 99, we can have a tens digit of 2, 3, 5, 6, 7, 8, 9 with a units digit of 0, 3, 5, 6, 7, 8, 9, giving 49 cases. Hence there are $6+9+49=64$ cases in 1–99. But when we wish to precede these with a hundreds digit, we have to also allow 00. Hence in the range 100–999, there are $6 * 65 = 390$ cases, so there are $390 + 64 = 454$ cases in the range 1–999. Any larger number has an O in it as it uses thousand or million or billion or .... The largest number we have described is indeed 999.

106. EVEN MORE SEQUENCES — 4

These are the numbers with no Is in their names. The last such is 777,777. The enumeration of these is complicated by the fact that three can occur as thirteen or thirty.

First we note that every number from 1,000,000 onward has an I in its name, so we consider numbers of the form *ABC,DEF*. Basically each digit can take on the six values 0, 1, 2, 3, 4, 7. But the digits *B* and *E* cannot take on the value 3 as this would produce thirty in the name. So far this permits 6*5*6*6*5*6 = 32400 cases. Now we must exclude the cases where thirteen occurs. This happens if and only if $BC = 13$ or $EF = 13$. There are 6*6*5*6 = 1080 numbers of the form $A13, DEF$. Similarly there are 1080 numbers of the form $ABC, D13$. However we have included the 6*6 = 36 numbers of the form $A13$, $D13$ in both countings. So the number of numbers in the range 000,000–999,999 with no Is is $32400 - 1080 - 1080 + 36 = 30276$. If we want to start at 1, as is usual in this type of problem, then we get 30275.

107. AN ELEMENTARY PROBLEM

They are all the single letter abbreviations of elements. Argon, Boron, Carbon, Fluorine, Hydrogen, Iodine, Potassium (from the Latin Kalium), Nitrogen, Oxygen, Phosphorus, Sulphur (or Sulfur), Uranium, Vanadium, Tungsten (= Wolfram), Yttrium.

I used this problem on the BBC Radio 4 program Puzzle Panel in 1999 and we received complaints that the symbol for Argon is Ar. However, I checked my 1955–1956 *Handbook of Chemistry and Physics* (bought when I was a high school chemistry student) and the symbol there is A. Later information (now on Wikipedia) is that the symbol was changed by the appropriate international commission on chemical nomenclature in 1957, but I have not yet found the actual decision, nor any motivation for it. I have found A still used in a *Pears Cyclopedia* of 1966. Any information about the reason for this change will be appreciated.

Chapter 8

108. STRANGE RELATIONSHIPS IN MUCH PUZZLING

There are several ways to approach this. The simplest is to observe that the baker's wife cannot be Mrs. Brewer (unless she talks to herself). She also cannot be Mrs. Baker, since the baker is not Mr. Baker. Hence she must be Mrs. Butcher. Now the brewer cannot be Mr. Brewer and we have just seen that he cannot be Mr. Butcher, so he must be Mr. Baker and the butcher must be Mr. Brewer. Mr. Brewer, the butcher, didn't marry his sister, Miss Brewer, nor did he marry Miss Butcher, so he must have married Miss Baker.

More generally, one can form a tabular arrangement of the surnames, jobs and wife's maiden names. One finds that there are just two arrangements compatible with the baker's wife's information. The fact that the baker's wife is not Mrs. Brewer determines one of these arrangements. Indeed, any such piece of information determines the entire arrangement.

If the baker's wife does talk to herself, then she is Mrs. Brewer and the other arrangement holds, in which Mr. Baker, the butcher, married Miss Brewer.

109. A HOLLYWOOD MURDER

In the first situation, we can proceed systematically. Alice's lie means that she doesn't know who did it, so it could be Benny, Carol or Donald. Benny's lie means that it is Alice, Carol or Donald. Carol's lie also means that Alice, Carol or Donald did it. Donald's lie means that Benny or Carol did it. The only person to fit all lies is Carol.

The second situation can be systematically attacked by considering just Alice lying, then just Benny lying, etc. But it's easier to notice that Benny and Carol are saying the same thing, so they are both lying or both telling the truth. Since there is just one liar, they must be telling the truth and so Benny is the killer. Donald is then the only liar.

110. THE DERANGED SECRETARY

Let's start off with a few letters and envelopes to explore the situation. When there is only one letter, Ms. Flubbit can't make a mistake and there is no problem. When there are two letters, there is only one way she can make a mistake and so I know that each envelope contains the other letter — without having to open any at all!

With three letters, A, B, C, Ms. Flubbit could have put them into envelopes A, B, C in just two possible orders: B, C, A or C, A, B. So I see that I must open at least one envelope. Let's open envelope A. It must contain letter B or C and this completely determines the contents of the other two envelopes. From this case, it is clear that we cannot deduce the contents when there are three unopened envelopes remaining. From the cases with 2 and 3 letters, it seems that we might be able to deduce the contents when there are two unopened envelopes remaining. But suppose we have envelopes A and B remaining and we know that they contain letters C and D. Then we cannot deduce which letter is in which envelope, and so we need to open another envelope. But with three letters, this didn't happen. Let us look at that case more carefully. When I open envelope A, I find letter B or C. If I find letter B, then I have two remaining envelopes B and C which contain letters A and C. Since letter C can't be in envelope C, it must be in envelope B and letter A must be in C. What has happened is that when we got down to two remaining unopened envelopes, the letter belonging to one of these last envelopes was one of the two remaining letters, hence had to be in the other envelope.

Now this situation does not always happen, as we saw by considering envelopes A and B with letters C and D. But can we make it always happen by some strategy? The answer is 'Yes' and it uses a simple technique. Instead of opening all but two envelopes at random, I start off with some envelope, say A. I look at the letter inside, which is addressed to someone that I'll call B. I search for the envelope addressed to B and open it to find a letter addressed to C. I put the envelope by the letter to B and search for envelope C, ... In the process, I may find the letter to A, but then I just start over again. When I have opened all but two envelopes, I have at most one opened envelope and one opened letter which are not paired off. When there is one of each, then we are in the same situation as having three letters and having just opened one of them. If there are no unpaired letters and envelopes, then we are in the situation of two letters, neither yet opened. In either case, we know what is in each of the two remaining envelopes.

111. IN THE PAWNSHOP

I'm sure you all decided that the poor sap of a friend lost. He had to pay $75 more to get the $100 back from Uncle George, so he wound up paying $150 for $100. He won't make that mistake again. But you've all made a mistake that does you credit. Apparently you've never had to pawn anything. Uncle George doesn't store your stuff and loan you money for free — there's interest. My local pawnbroker says it's currently 4% per month or part of a month. Since the $100 bill was only there for a few days, the interest on the $75 loan was $3 and the sap had to pay $78 to redeem the bill.

[This was a favorite puzzle of Lewis Carroll. I have no earlier reference to it.]

[In July 1997, I saw an article saying the interest rate is 6% per month, simple interest, for six months, then the item goes on sale, usually by auction. The pawnbroker keeps 136% of the amount of the loan and the rest goes to the pawner.]

112. A TAXING ROAD PROBLEM

Consider the road from the highway up to Able. All four of them will use it, so they should each pay a quarter of the cost of that mile. I.e. they should all contribute $300 toward the first mile. But the second mile is only used by Baker, Charlie and Dog, so they should each pay $400 toward the second mile. Similarly Charlie and Dog should each pay $600 toward the third mile and Dog should pay $1200 for the last mile. Then Able pays a total of $300, Baker — $700, Charlie —$1300 and Dog — $2500.

This proposal may not strike everyone as reasonable. In fact Dog threatens to pull out — he says he'll wait till the road gets to Charlie's and then he'll only have to pay $1200 for the last mile. Charlie wonders whether Able might use the road to go up as well as down, though Able denies that he has any business up the road.

Problems of this sort are surprisingly old. In the *Patiganita* of Sridhara (900), we find a troupe of religious dancers performing a day-long service for four men. The first man leaves after a quarter of the day, the second man after half the day and the third man after three quarters of the day. Here the cost is divided in the proportion 3 : 7 : 13 : 25 which we found above and this seems beyond dispute.

113. TRUE OR FALSE?

Your first responses should be: “I come from Truth and this is Truth”. If you are on Truth, you are speaking the truth and they will accept you as a Truthteller. If you are on Falsehood, then you are lying and they will accept you as a Liar.

Your question is actually much easier, though the complexities of the situation are designed to make you try for very complex questions. Any simple question with an objective answer will do, e.g. “Do you live on this island?” or “Is the sun shining?” (Of course, the latter question assumes that it is clear whether the sun is shining or not.)

[Despite their simplicity, truthtellers and liars problems only seem to date back to about 1930. The earliest example I know of is June 1929 when it was given by Nelson Goodman in the Brainteasers column of *The Boston Post*. I found this described in a letter from Goodman to Martin Gardner, apparently in the 1960s, where he says he ‘made it up out of whole logical cloth’ and submitted it to the paper. I am always interested to hear of early occurrences of problems such as this one.

The second part of the above problem is adapted from one in Peter Eldin, *Amaze and Amuse Your Friends*, 1973.]

114. HOLMES VERSUS LESTRADE

Holmes asks Xenia: “Did Yolanda do it?” If Xenia is the guilty party, then she will answer yes or no as she wishes. But if she is innocent, she will tell the truth. Hence if she answers yes, then the guilty party is either Xenia or Yolanda, while if she answers no, then the guilty party is either Xenia or Zelda. In either case, Holmes knows an innocent party and can determine the guilty party by asking the innocent party if Xenia did it.

There are many other ways of organizing the questions. It is pretty clear that one yes or no question cannot determine the murderer among three suspects, but I do not know if one can find the murderer with two questions and suspects set in advance — Holmes used the answer to the first question to determine whom to interrogate next, although his questions were fixed in advance. Lestrade was probably thinking of asking Xenia if Yolanda did it, then asking Yolanda if Zelda did it and finally asking Zelda if Xenia did it.

[Adapted from F. W. Sinden; Logic Puzzles; *Studies in Mathematics* (School Mathematics Study Group) XVIII (1968) 197–201.]

115. A STRIKING PROBLEM

I could have been awake for an hour and a half and it could then be either 1:30 or 2:00. This can happen in two ways.

I could have woken up as I heard one stroke. Not knowing whether it was part of an hour or not, it could be any hour or half-hour. After half an hour, I hear one stroke. Then I know it is either a half-hour or 1:00. After a second half-hour, I hear a single stroke again. Then I know it must be either 1:00 or 1:30, but I won't know which until I hear the clock after another half-hour — if I then hear a fourth consecutive single stroke, it is 1:30, while if I then hear two strokes, it is 2:00.

Alternatively, I might have woken up just after the clock struck, without hearing it. After half an hour, I hear one stroke and the situation is the same as the above. However, in this case, I do not get to hear four consecutive single strokes and I have waited perhaps a second less than in the previous case.

[Adapted from Sam Loyd Jr.; *Sam Loyd and His Puzzles*, 1928.]

116. A HAIRY SITUATION

I get either 2 or 238, depending on how one interprets the problem.

Suppose there are N hairs on the head of the hairiest inhabitant. It is clear that N is fairly large, certainly greater than 237. Then we have $N+237$ inhabitants who can have $1, 2, \dots, N$ hairs on their heads. If the first $N-1$ inhabitants have $1, 2, \dots, N-1$ hairs and the last 238 inhabitants have N hairs, then there are 238 persons with the same number of hairs.

A different arrangement would be to have the first $N-237$ inhabitants having $1, 2, \dots, N-237$ hairs. Then the remaining 474 persons could be paired off, two each having $N-236$, $N-235, \dots, N$ hairs. Then there are at most 2 people with the same number of hairs. But there are 474 people who have the same number of hairs as someone else. This must be what Pearson had in mind — but the previous case shows that the solution of this reading of the problem can be reduced to 238!

117. IN THE DARK

In the first case, I need 3 socks to be sure of having a pair of socks of the same color, but I need 11 gloves since I might pick up all 10 right gloves before getting a left, making a total of 14 items.

The second case is a bit trickier. I can pick 12 socks to be sure of getting pairs of both colors and 11 gloves to be sure of getting a pair, making a total of 23 items. But I can pick 16 gloves to be sure of getting pairs of both colors and 3 socks to be sure of getting a pair, making a total of only 19 items.

118. STRANGE RELATIONSHIPS IN LESS PUZZLING

Suppose Mr. Smith married Miss Jones. Then if Mr. Jones married Miss Smith, Mr. Robinson would have had to marry his own daughter, which is not permitted. So Mr. Jones married Miss Robinson and Mr. Robinson married Miss Smith. So Mr. Smith's father-in-law is Mr. Jones, whose father-in-law is Mr. Robinson, who married Miss Smith.

Suppose Mr. Smith married Miss Robinson. Then Mr. Robinson must have married Miss Jones and Mr. Jones married Miss Smith. Then Mr. Smith's father-in-law is Mr. Robinson, whose father-in-law is Mr. Jones, who married Miss Smith.

The information given does not determine which of the two cases actually occurred, but the question has the same answer — Miss Smith — in either case!

119. FERRYING FOUR JEALOUS COUPLES

Let the couples be A, a, B, b, C, c, D, d and suppose we are traveling from the left bank (L) to the right bank (R). The number of people at R is changed by at most one by each back and forth trip. After some back and forth crossings, we will have three people on R. Because of the jealousy conditions, these can only be three women. Consider the last time this occurs. Then the next back and forth crossing must bring two people to R and send one back to L in order to get four people on R. But we cannot send either two men or a couple across without having an unescorted woman on R and there is only one woman left on L so we cannot send two women. Hence there is no way to increment three persons at R to four persons by back and forth trips.

For four couples and an island (I), with no bank to bank crossings:

Send ab from L to I and b back to L.
Send bc from L to I and c back to L.
Send AB from L to I, send ab on from I to R and b back to I.
Send AB from I to R and B back to I, then back to L.
Send cd from L to I and d back to L.
Send BC from L to I, then on to R, and a back to I.
Send ab from I to R and C back to I and then back to L.
Send CD from L to I, then on to R, and b back to I.
Send bc from I to R and b back to I, then back to L.
Send bd from L to I and then on to R.

This uses 26 trips and is minimal because the fastest possible ferrying in a two person boat occurs when every L to I or I to R trip has two people aboard and every R to I or I to L trip has one person, as in this solution. It takes at least 13 crossings to transport 8 people across a stretch of water, even with no jealousy conditions, and we have two stretches of water, because bank to bank crossings are not allowed.

When bank to bank crossings are allowed, a computer search by my student Ian Pressman turned up a 16-trip solution, which we showed to be minimal.

Send ab from L to R and b back to L.
Send bc from L to I and c back to L.
Send AB from L to R and B back to L.
Send Cc from L to R, c back from R to I and b from I to L.
Send Bb from L to R and C back to L.
Send CD from L to R and a back to L.
Send ad from L to R, a back from R to I and ac from I to R.

(There is a question whether the fourth passage is satisfactory — might Cc decide to stop on the island? We do not know if this question can be avoided.)

120. FERRYING A FAMILY

Child 1 takes 3 across, then $5, 7, \dots$, until all the children, with odd numbers greater than 1, are across. The father takes 1 across and returns. The father takes 2 across and sends 1 back. Child 1 now takes child 4 across, then $6, 8, \dots$, until all the even children are across — finally 1 stays across when he/she brings the last even child across.

Every child except 1, 2 and the last even child is taken across by 1 and then 1 returns, so this takes $2(N-3) = 2N-6$ crossings. Then we must add the four crossings in the middle and the final crossing of 1 with the last even child, making a total of $2N-1$ crossings. This is actually the minimal number of crossings to get $N+1$ people across a river even when there are no restrictions.

121. HAPPIER FAMILIES IN MUCH PUZZLING

In the previous problem of this sort, with three widowers and daughters, there were just two possible ways the weddings could have taken place and we considered both cases. Here there are actually 9 ways and it would be tedious to examine all of them. Instead, let us consider the step-son function on the four men. Because this is a permutation of the four men, if we keep iterating it, we must eventually come back to where we started, i.e. the process must cycle from any starting man. If we take all the distinct cycles, then every man occurs just once in these cycles, so the lengths of the cycles must add up to four. Since no man can marry his own mother, there are no cycles of length 1. Consequently, there are no cycles of length 3 and only two situations can occur — either there is a single cycle of length 4 or there are two cycles of length 2. As an example of the first case, we might have Archer marrying the widow Baker, Baker marrying the widow Cobbler, Cobbler marrying the widow Dyer and Dyer marrying the widow Archer — there are 6 such cycles corresponding to the 6 permutations of Baker, Cobbler, Dyer. As an example of the second case, we might have Archer marrying the widow Baker, Baker marrying the widow Archer, Cobbler marrying the widow Dyer and Dyer marrying the widow Cobbler — there are 3 such patterns corresponding to whom Archer marries. Since 2 divides 4, in either situation, the fourth step-son of every man is himself!

[Similar analysis could have been done in the three-couple case, where the only possibilities are cycles of length three. With five couples, things will not be so simple!]

122. FOUR GREAT-GRANDPARENTS

My first solution was to consider two married couples, say $A + B$ and $C + D$. Both of these have children, let us say a son E and a daughter F. Both couples then divorce and they remarry, giving us couples $A + D$ and $C + B$, who produce a son G and a daughter H. Now E and F marry, producing a son I, and G and H marry, producing a daughter J. E and F are both half-siblings to both of G and H, so I and J are sort of 'double half-cousins'. If they marry, which is probably permitted, their offspring will have only four great-grandparents.

The second method is actually a bit simpler. Start with two married couples, $A + B$ and $C + D$, and let the first produce two sons, E, F, while the second produces two daughters, G, H. Now the brothers marry the sisters, giving couples $E + G$, $F + H$, producing a son I and a daughter J. Again, I and J are sort of 'double cousins'. If they marry, which may be permitted, their offspring will likewise have only four great-grandparents. Currently I think the first case may be more likely than this case.

[At the time, I didn't know any actual case of this phenomenon, but William III's mother was Mary, the sister of James II who was the father of William's wife Mary (II), i.e. William and Mary were cousins, so their children would have had six great-grandparents.

I later found that Victoria and Albert were cousins, so all their children had six great-grandparents. Even later, I found that Prince Don Carlos of Spain (1545–1568) had only four great-grandparents — perhaps someone can provide details.]

Chapter 9

123. HOME IS THE HUNTER

This seems quite impossible, but is actually quite easy. Hiawatha is anywhere less than ten miles from the North Pole, but not at the pole. So when he heads north and goes ten miles, he crosses over the Pole and continues a bit beyond. After lunch, he returns the way he came.

[I invented this around 1980 and have seen it used in several books.]

124. THE STABLE TABLE FABLE

Consider the table as placed. One foot is up in the air. (Actually the table will rock with just two diagonally located feet touching the ground, but consider it as having settled with three feet on the ground.) Label this foot as A and let the other feet be B, C, D, in order. Imagine rotating the table by 90° so that foot B is now where A was, keeping feet C and D on the ground as you turn. By the symmetry of the table, the 'wobble' at A (i.e. the distance between foot A and the ground) is now the wobble at B and foot A is now on the ground. At some point in the rotation, the wobble of A must have become zero and at this point B has no wobble, so this is the desired position.

More carefully, let a and b denote the wobbles at A and B. Let $a_0 > 0$ be the initial wobble at A. Consider the difference $a - b$. In the initial position, this has value $a_0 > 0$ and in the final position, it has value $-a_0 < 0$. Hence it must be 0 at some intermediate position, i.e. $a = b$ there. But $a = b > 0$ is impossible — you can't have two adjacent feet in the air at once. So $a = b = 0$ at this position.

[Mathematical readers will recognize that the proof uses the Intermediate Value Theorem for continuous functions. This is one of the basic properties of continuous functions and one that is intuitively plausible to the general reader. I have assumed that my patio has a continuous surface, which is not unreasonable, though my friends don't think much of my work! I have also assumed that the table feet are points, which is only approximately true. In November 2014, I heard that this solution depends on the ground not being too irregular, but I didn't get details — I'd be happy for enlightenment.]

125. MATCHSTICK TETROMINO QUADRISECTION

The connected areas formed by 4 of our unit squares are the following, called tetrominoes.

All of these, except the square, have 10 boundary segments, while the square has 8. When 4 of these shapes are formed in our 4 by 4 square, there are 16 external segments (which are already drawn) and n internal segments (which are our matchsticks). The external segments belong to one tetromino, while the internal segments will belong to two tetrominoes. Hence the four tetrominoes will have $16 + 2n$ segments in total. This number must equal one of the following sums:

(a) $8 + 8 + 8 + 8 = 32$;
(b) $8 + 8 + 8 + 10 = 34$;
(c) $8 + 8 + 10 + 10 = 36$;
(d) $8 + 10 + 10 + 10 = 38$;
(e) $10 + 10 + 10 + 10 = 40$.

Hence n must be 8, 9, 10, 11 or 12.

For $n = 8$ matchsticks, we are in case (a) and there is only one way to make such a pattern. For $n = 9$, we must have three square tetrominoes and one other. But there is no way to put three square tetrominoes in our 4 by 4 region which leaves any connected region other than another square tetromino. So $n = 9$ is impossible. The case $n = 10$ uses two squares and I find three distinct ways to make such a pattern. The case $n = 11$ uses one square. I find four distinct patterns, all using two L pieces and one I piece. The case $n = 12$ has many solutions — I didn't find them all. So our problem's answer is 8, 10, 11 or 12.

126. OFF THE RAILS

Let the ends of the rail be at A and B, with the center at C. When the center is lifted to C', the ends move in to positions A', B', as indicated in the sketch below.

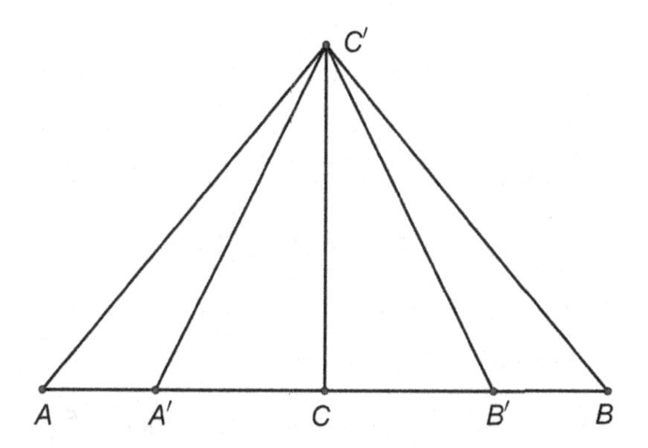

Then $A'CC'$ is a right triangle (assuming the rail remains straight) with $A'C' = 2640$ feet and $C'C = 200$ feet. Hence $(A'C)^2 = 2640^2 - 200^2 = 6929600$, so $A'C = 2632.41\ldots$ feet. Thus $AA' = 2640 - A'C = 7.59$ feet and the two ends come together by twice this amount, i.e. by 15.17 feet or 15 feet 2.08 inches.

Hence Jonathan Always is correct. If we read Ripley as saying AA' is about 3 inches, i.e. .25 feet, then $(C'C)^2 = 2640^2 - 2639.75^2 = 1319.94$, so $C'C = 36.33$ feet. I am unable to figure out what Ripley intended but $C'C$ is about a yard.

127. LEANING TOWERS OF NEW YORK?

Let A, B be the bases of the towers and let A′, B′ be the tops. Let O be the center of the earth. Let d be the distance between A and B, D be the distance between A′ and B′, h be the height of the towers and R be the radius of the earth, as shown in the diagram.

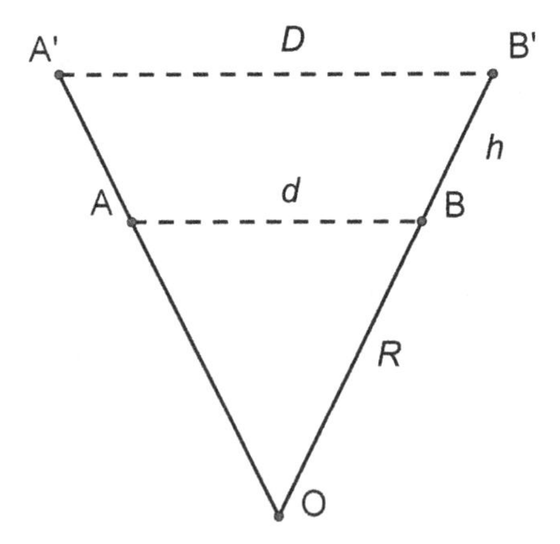

By similar triangles, or similar sectors of a circle, we have

$$D/d = (R + h)/R.$$

Set $e = D - d$. Then $(d + e)/d = (R + h)/R$, so that $e/d = h/R$ and $h = eR/d$. Using our numbers, we have

$$h = (1\ 5/8)(4000/1) = 6500 \text{ inches} = 541.67 \text{ feet} = 541 \text{ feet } 8 \text{ inches}.$$

[I read these numbers some time ago, and I don't know how accurate they are.]

128. SQUARE CUTTING

Fold the square along a diagonal. Fold the resulting triangle along the bisector of its right angle, which is along the other diagonal of the original square. This yields another right triangle having both the vertical and horizontal midlines of the square lying along the bisector of the right angle of the triangle. Hence a single cut along this bisector will produce four squares.

To divide into four triangles, we fold the square to make the diagonals lie together, rather than the midlines. This is done by simply folding along one midline, then the other, to yield a square and then we cut along the diagonal which passes through the corner which was the original center of the square.

[With a bit more folding, one can divide the square into 9 squares with a single cut, and indeed into mn rectangles. I discovered this some years ago. Martin Gardner told me that it is an old magician's trick, but I had never seen it in print until very recently when I found it in a 1940 German magic book.]

129. A SIMPLER DISSECTION

The first solution is shown below.

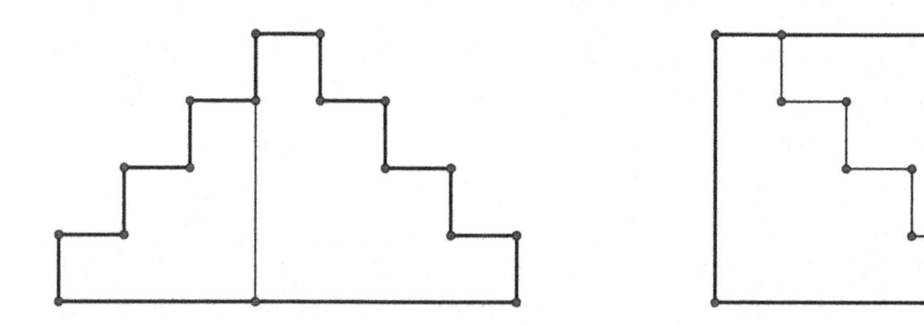

The second solution is shown below.

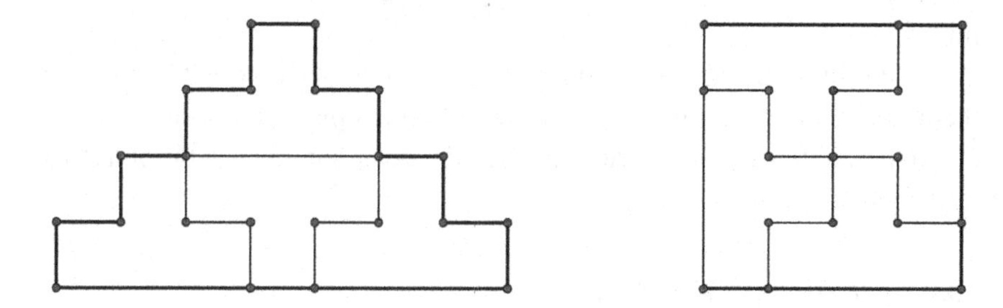

130. CUTTING THE CHRISTMAS CAKE CORRECTLY

The solution is really quite easy. Simply divide the perimeter of the cake into 5 equal parts and cut from the center to these divisions. It is easier to consider one side divided into five equal parts. Looking down on the cake, we see that the cuts give five triangles of the same altitude and the same base. So the volume of cake is the same in each part and the amount of top frosting is the same and the amount of side frosting is the same. For the division of the whole cake, each person's portion comprises four such triangular pieces — but they may extend around a corner so it is not so easy to see the equality of the portions.

[We have assumed the frosting is of negligible thickness. This is not true for Christmas cakes. The method also works for thick frosting though the geometry is a bit harder.]

131. RETRIEVE THE BOAT

One person holds an end of the rope at the edge of the pond — or it can be tied to any convenient object. Someone then takes the other end and walks around the pond, keeping the rope fairly taut. The rope will catch the boat and bring it to shore. One needs enough rope to reach across the pond, which is guaranteed by the problem stating that the rope would reach to the boat and back when the boat is in the center. One may object that the rope might sink. Most ropes are not much denser than water and some are lighter, but if this seems to bother you, you can tie some of the sticks near the middle of the rope so the rope will catch the boat and not pass under it. If the rope is light enough, the two people holding it may be able to suspend it so it's just touching the water in the middle.

[I am sure one can think of other methods, though I've tried to eliminate them. On Clapham Common, there are often people flying kites and one could use a kite to carry a string to snare the boat, but this is a bit of a *deus ex machina* solution.]

132. BRIDGING THE MOAT

For the square case, place the first board diagonally across a corner of the moat. The center of this board is now distance $L/2$ from the corner of the moat. Placing the end of the second board at this center will give a walkway to the island if $L+L/2 \geq \sqrt{2}$, i.e. if $L \geq 2\sqrt{2}/3 = .9428$. Since L is only a few percent shorter than 1, L is enough larger than .9428 to allow for the board widths which have been ignored in the above calculations.

For the circular case, place the first board as a chord of the circular moat. Its center is at distance s from the island, where $(r+s)^2+(L/2)^2 = (r+1)^2$.

This yields $s = -r + \sqrt{[(r+1)^2 - (L/2)^2]}$, and we will have a bridge if $s \leq L$. For $r = 0$, $s = \sqrt{[1 - L^2/4]}$ and this is $\leq L$ if and only if $1 - L^2/4 \leq L^2$ or $L \geq 2/\sqrt{5} = .8944$, so that considerably shorter boards can be used here. When $r = 1$, $s = -1 + \sqrt{[4 - L^2/4]}$ and this is $\leq L$ if and only if $4 - L^2/4 \leq (L+1)^2 = L^2 + 2L + 1$. Rearranging gives the equivalent expression $5L^2/4 + 2L - 3 \geq 0$, which holds for positive L greater than or equal to the positive root of the quadratic expression, i.e. for $L \geq (-4 + 2\sqrt{19})/5 = .9435$.

[The extra credit problem is more easily done for the circular case. Let $R = r + 1$ be the radius of the outer edge of the moat. Place some boards as chords of this circle. The centers of these boards are at radius R_1, where $R_1^2 = R^2 - L^2/4$ from the center of the island. By using enough overlapping chords, we effectively reduce the moat to a circle of radius R_1. We repeat the process, which reduces the moat to radius R_2, where $R_2^2 = R_1^2 - L^2/4 = R^2 - 2L^2/4$. In general, n repetitions of the process reduce the square of the radius to $R^2 - nL^2/4$, which eventually becomes zero.]

133. FIND THE CENTER

Remove the top sheet from the pad. Place the pad on the circle so the corner actually touches the circle. Since the angle in a semicircle is a right angle and conversely, the two edges of the pad cross the circle at the ends of a diameter. Mark these with the pencil and then use the edge of the pad to draw this diameter. Repeat for another diameter. The intersection of the diameters is the center.

You may feel this is a little sneaky, but I did say you could use the pad. Alternatively, you can fold over a corner of the page to touch the circle and then fold over an edge to let you draw the diameter, which solves the problem without even using the pad!

134. A FALSE CUT

What's wrong with the solution is that when one piece is slid up and over, the result is not a square, but is a 3×6 rectangle with two extra unit squares sticking out. If this method of cutting is going to work, the original piece cannot be square, but must be a 7×5 rectangle and then the cut shown below allows you to reform it into a 5×6 rectangle.

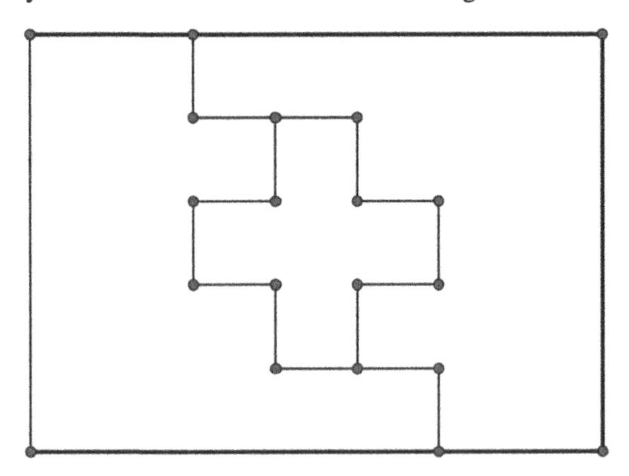

135. THE SPIDER SPIED HER

To visualize the shortest route, imagine the cylinder cut lengthwise diametrically opposite to the spider and then spread out flat. We then have a rectangle $6''$ high by $30''$ long, with the spider along the midline and $6''$ from an end, say the right end. To get to the fly, the spider must go to an end of the pipe (i.e. to an end of our rectangle) and then along the outside of the pipe. To see both parts of this route, we consider our rectangle reflected in its right end. Then the shortest distance will be a straight line from the spider to the reflection of the fly. If the fly is diametrically opposite the spider, it must be considered as lying on both the top and bottom edge of our rectangle and the spider has two (at least — see below) symmetric shortest routes to it. Let v denote the difference in the vertical positions of the spider and the fly in our rectangle (i.e. how far the fly is from the horizontal midline) and suppose the fly is f inches from the end that the spider is $6''$ from. If the spider goes to the near end, then his shortest route will have distance d_1 such that $d_1^2 = v^2 + (f+6)^2$. But if the spider goes to the far end, then his shortest distance d_2 satisfies $d_2^2 = v^2 + (24+30-f)^2$.

These two expressions are equal if and only if $f = 24$.

One can see this without even using the theorem of Pythagoras. Since the vertical components are equal, the distances are equal if and only if the horizontal components are equal, i.e. $f + 6 = 54 - f$ or $f = 24$. Then the fly is as far from the far end as the spider is from the near end.

If the fly is 24″ from the spider's end and is diametrically opposite, then there are four shortest routes for the spider.

136. SNOOKERED!

Dropping perpendiculars from the centers A, B, of two corner balls to the rack, we see that the inside edge of the rack consists of an inner part of length $8r$ and two end parts of length CD.

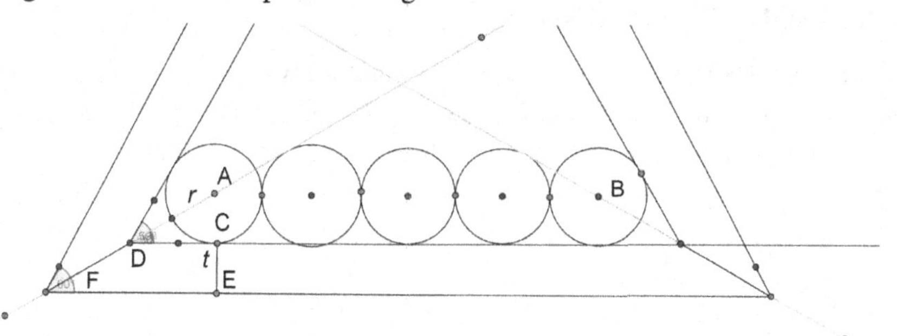

Since $\angle CAD = 60°$, we have that $CD = \sqrt{3}$ CA, i.e. CD= $r\sqrt{3}$, so the inside edge length is $(8 + 2\sqrt{3})r = 11.46r$. The outside edge length is $8r$ plus two parts of length EF, which is just $\sqrt{3}(r + t)$, so the outside edge length is $8r + 2\sqrt{3}(r + t)$.

137. GOAT AND COMPASSES

Let the shed have dimensions $2a$ by $2b$ (this avoids fractions later on). Then the length of the rope is $2a + 2b$. Suppose the stake is distance x from the middle of side $2a$, where $x \leq a$. Then the goat can graze over:

(1) a semicircle of radius $2a + 2b$;
(2) two quarter circles of radii $2b + a + x$ and $2b + a - x$;
(3) two quarter circles of radii $a + x$ and $a - x$.

Adding these areas together, we find the area grazed is

$$3\pi a^2 + 6\pi ab + 4\pi b^2 + \pi x^2 = 3\pi(a + b)^2 + \pi b^2 + \pi x^2.$$

This is minimized when $x = 0$, so we don't want to be in the middle of a side. The maximum occurs when x is maximal, i.e. for $x = a$, so we want to be at a corner — and all corners are the same — giving an area of $4\pi a^2 + 6\pi ab + 4\pi b^2$. The minimum occurs when we are in the middle of the longer side, so that $b \leq a$.

[The use of x as the distance from the middle exploits the symmetry of the problem to obtain a simple formula for the area, which is symmetric in x. If one takes the distance from the corner, the expression is more complicated.

The Goat and Compasses is an occasional pub name in England, apparently a corruption of the religious phrase 'God encompasseth'].

138. SQUARE BASHING

Suppose the points are labeled A, B, C, D and we want opposite sides of the square to pass through points A and C and B and D. Then the line segment AC is a transversal between the lines through A and C and making a certain angle a with these lines. A transversal between the lines through B and D making the same angle a with these lines must have the same length as AC and is perpendicular to AC. So first draw AC. Then draw the line perpendicular to AC and through B. Mark off the distance AC along this line to get a point E, so that BE = AC. Then E must lie on the side of the square which passes through D, so the line DE must lie along one side of the square. Constructing parallels and perpendiculars to DE gives us the other sides. If E = D, we can choose the first line arbitrarily through D. In general, we can take E on either side of B to get two points, say E and E′, giving two solutions. But when E = D, the solution for E′ is degenerate with a square of side 0. We can generally find six solutions by choosing to put points A and B or A and C or A and D on opposite sides of the square.

139. SHADES OF EUCLID!

The solution uses similar triangles, and is pretty easy — once you make a drawing.

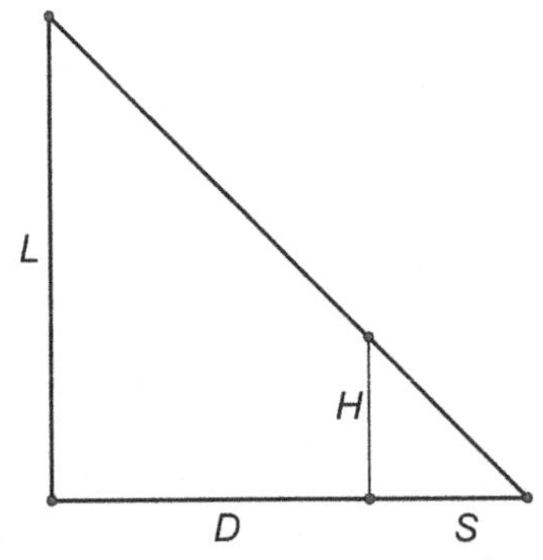

Let L be the height of the lamp and H the height of the person and let the person be at distance D from the lamp. Then the length of the person's shadow is S, where $S/H = (S + D)/L$ by similar triangles. This is readily solved for S, getting $S = HD/(L - H)$. Since S is directly proportional to D, the rate of growth of the shadow is in the same proportion to the speed of the person. For example, if $L = 3H$, then the shadow's length is growing at 1/2 the speed of the person. So the shadow may actually be growing slower than the person is moving, and it is not growing faster and faster.

Some may feel that the correct distance to consider is not S but $S + D$, which is the distance of the shadow's head from the lamp post. We have $S + D = LD/(L - H)$, so this distance increases faster than the speed of the person (since $L > H$), but it still does not go 'faster and faster'.

140. MATCHSTICK SQUARES

With all the careful phrasing of the problem, very few people realize that the size of the squares is not specified. If we make the side of the square be half the length of the match, one can put together the matches somewhat like making a house of cards in the pattern shown below.

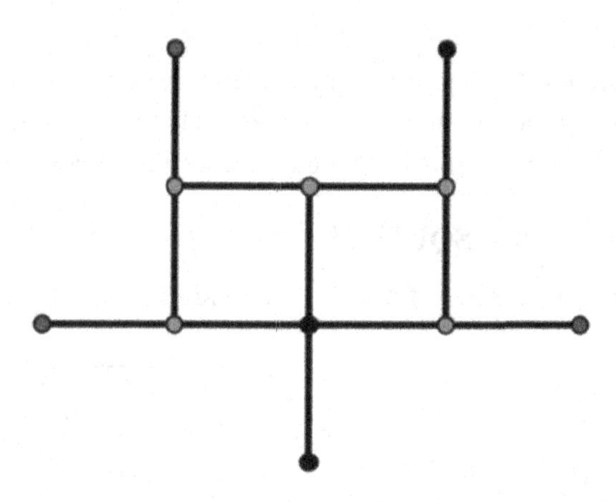

Research questions: I believe the number of squares that can be formed from n matchsticks, where the sides of the squares are equal to a matchstick, has been determined, but I doubt whether anyone has investigated the present case where the sides of the squares are equal to half a matchstick.

141. CUTTING UP A SQUARE AGAIN

First, it helps to figure out what the triangles must look like. They have altitude equal to the side of the square and base equal to 2/3 of the side of the square, and hence area 1/3 of the square. To form them, we could make a zig-zag cut which starts at the lower left corner, goes to a point 1/3 of the way along the top edge, then back to the point 2/3 of the way along the bottom edge, then back to the upper right corner. This gives four pieces — two of these fit together to form a triangle identical to the other two. (Sneaky, or what!)

Now we must figure out how to do the operation in one cut. If we fold the square along the verticals through the 1/3 and 2/3 points, we get a triple thick rectangle and we simply cut along one of its diagonals to achieve the desired result.

142. CUTTING UP A TRIANGLE

Again, it helps to work out the size of the resulting triangles. If our original triangle has side A, then the new triangles must have side $A/\sqrt{3}$. This indicates that the new side must be related to the altitude of the original triangle, whose length is $A\sqrt{3}/2$. The new side is just 2/3 of this length and this is well known to be the distance from a vertex to the point (the centroid) where the three altitudes meet. This should give us the clue that we cut the triangle along all three altitudes. This gives us six pieces, but they fit together in pairs to form three equilateral triangles.

143. MAKE THREE SQUARES

Arrange the nine matches to form the following

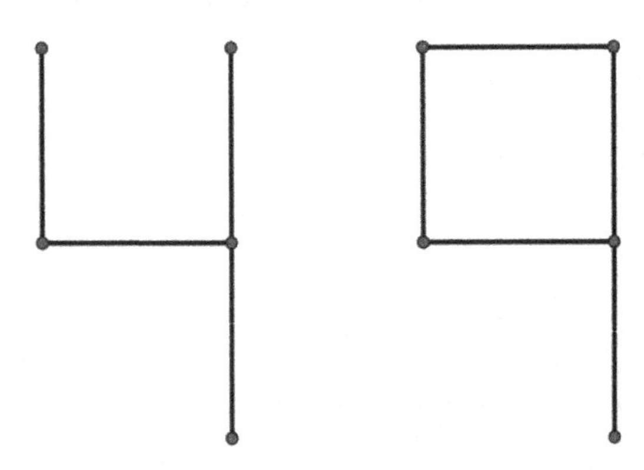

Then 4 is a square, 9 is a square and 49 is a square. This is the idea I had when I made up the puzzle, but various friends gave more examples and then I found another solution which another friend produced later.

Solution 2. Use the matches to form a triangular prism. One may object that this also makes two triangles.

Solution 3. Make three squares forming three faces of a cube, all meeting at one corner.

Solution 4. Make two adjacent squares with seven of the matches. Now bisect each of the squares with a match parallel to the common edge of the squares. This produces a row of four adjacent half-squares as below. The middle two form a new square. Here one may object that the squares are overlapping.

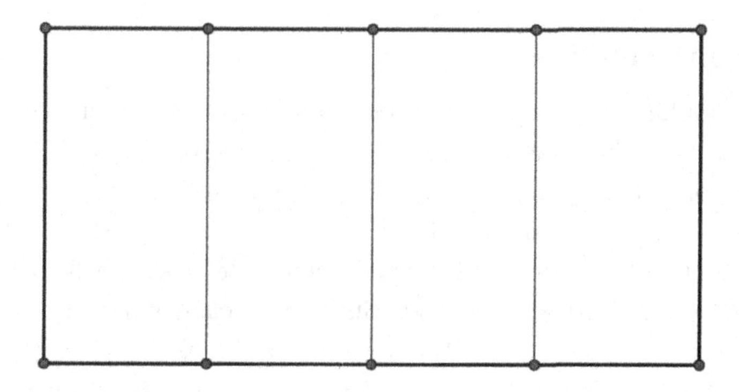

Solution 5. Use the matches to make the figures 0, 1 and 4.

One can use the matches to make squares whose edge is half the match length, but one only needs eight matches to make three squares.

There are other solutions which use the fact that matches have squared off ends and have square cross-section, but these properties do not hold for paper matches torn from a matchbook or for other equivalent objects like toothpicks and hence I don't consider them quite reasonable.

There are probably more solutions, perhaps bending the rules a bit more. For example, one can form 1 4 4 which gives four squares!

Chapter 10

144. TIME FLIES

One actually can arrive back in New York at an earlier hour than one leaves, but it is on the next day. We cross the International Date Line.

145. SAM'S ESTATE

Simon shouldn't buy. Looking in a dictionary or other reference tells us that a rod (= pole = perch) is 5 1/2 yards, a chain is 22 yards = 4 rods and a furlong is 220 yards = 40 rods. Hence the measurements of the tract are: $AC + BC = 70$ rods; $AB + BC = 110$ rods; $AB + AC = 100$ rods. Adding all these gives us twice the perimeter as 280 rods, so the perimeter, $AB + BC + AC$, is equal to 140 rods. Subtracting each of the measurements from this gives us: $AB = 70$, $AC = 30$, $BC = 40$. Now $AB = AC + BC$, which can only happen if and only if C is on the line between A and B. Hence there is no land at all in this tract! Unless you are in the pay of Slick Sam, you should advise Simon not to buy it.

146. FLAT EARTH?

A level sight from one end, call it A, of the canal gives a straight line tangent to the earth at A. The canal itself bends with the earth, giving a circular arc of length $d = 6$ miles. Let the other end of the canal be B and the center of the earth be C. The line CB will meet the tangent line of sight at a point B'.

The distance $B'B$ is the 'depression' of B as viewed from A. See the diagram.

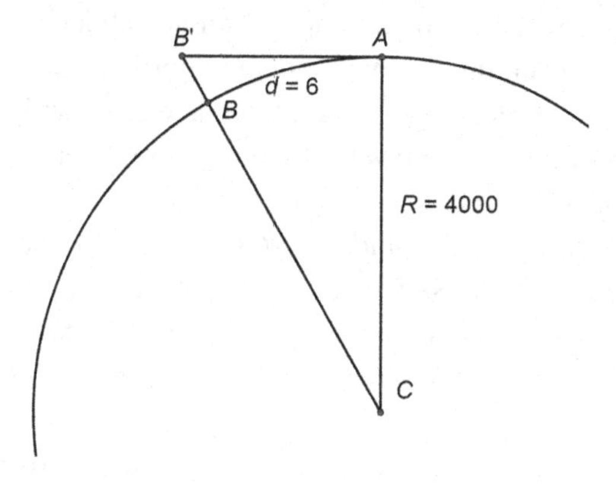

Now $AB'C$ is a right triangle with $AC = 4000$ miles. The distance AB' is close enough to AB that it is reasonable to set $AB' = d$. Then $B'C^2 = 4000^2 + d^2$ and $B'B = B'C - BC = \sqrt{(4000^2 + d^2)} - 4000$ miles. Since d is small, $(4000 + d^2/8000)^2 = 4000^2 + d^2 + d^4/4(4000)^2$ is approximately equal to $4000^2 + d^2$. That is, $\sqrt{(4000^2 + d^2)}$ is approximately $4000 + d^2/8000$ and so $B'B$ is approximately $d^2/8000$ miles, which is $7.92\, d^2$ inches. Thus Fort is reasonably correct. For $d = 6$ miles, we get a depression of 285.12 inches.

In the above, we have made two approximations, both of which are reasonable for small values of d. Using a little trigonometry, one can find the distance $B'B$ exactly. A 10 decimal place calculator gives 285.12 inches, so the approximations have no observable effect.

[The more accurate analysis is as follows. The angle ACB is $d/4000$ (in radians), which we denote by a. Then $B'B = 4000 \sec a - 4000 = 4000 (\sec a - 1)$ miles. Using the Maclaurin series for $\sec a$, we get $4000 (\sec a - 1) = 7.92\, d^2 + .00000020625\, d^4 + \cdots$ inches. For $d = 6$, the second term is 0.00004455 inches, which can reasonably be neglected in comparison to 285.12 inches.]

147. AN ALL-DAY SUNRISE

He could be telling the truth, if he started out near one of the poles. If one is a distance s from a pole, the parallel of latitude is a circle of radius

$r = R \sin s/R$ and $2\pi R \sin s/R$, where R is the radius of the earth. In 24 hours, Uncle Tall-Storey flew 9600 miles. Using $R = 4000$ miles, we find s is about 1568 miles, which corresponds to latitude about $67\frac{1}{2}^\circ$ either North or South. In the North, this is about a degree north of the Arctic Circle. If he started out 1568 miles from a pole, then he circumnavigated the globe in exactly 24 hours and the sun would seem to remain in the same position in the sky as when he started.

When he went east, he would go around the earth once and the earth would rotate once, so the sun would appear to go around twice. On an equinox, he would see the sun set after 6 hours, then rise after another 6 hours, set after another 6 hours and, finally, rising after a total of 24 hours. Thus he would have had two complete days in 24 hours! (The lengths of daylight and night-time would vary as the date moves away from an equinox, but he would still see two days and two nights except when the date is very close to a solstice.)

[More precisely, there is an additional effect due to the movement of the earth in its orbit which changes the time of sunrise from day to day, but this would not be enough to be noticeable unless the date is close to a solstice.]

148. A TALLER STORY

If Uncle goes twice as fast, then in 24 hours, he goes around the earth twice to the west, while the earth is going once to the east. This makes it appear as though the sun has gone around the earth once, but from west to east! So as soon as he starts at sunrise, the sun will set in the east behind him. After about 12 hours, he sees the sun rising in the west in front of him — as he reaches his starting point. If he continues, then after another 12 hours (making 24 hours in all), he again arrives back at his airstrip as the sun is setting in the east behind him. He lands and watches the sun rise in the east!

To get the same effect at 400 miles per hour, he would have to be on a parallel whose length is 4800 miles and this is at about 769 miles from the pole, which is at latitude about 89°.

[Again there is a additional effect due to the movement of the earth along its orbit, but this will only change some of the times and not the qualitative effects, provided we are not too close to a solstice.]

149. A POINT WITH A VIEW

For convenience, imagine we are viewing the earth from a point A at a height h directly over the North Pole P. Let C be the center of the earth and let R be its radius. Draw a tangent line from A to the earth, meeting the earth at B, so angle ABC is 90° and $BC = R$. See the diagram.

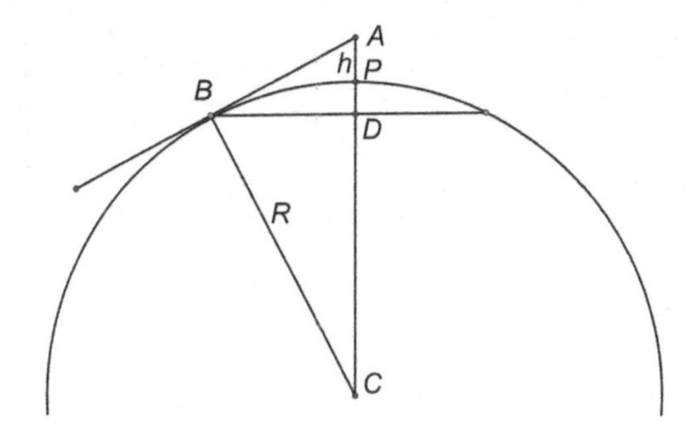

The part of the earth visible is that contained between the North Pole and the plane through the circle of latitude at B. Let D be the point where this plane cuts the line APC. Then the amount of area visible will be $PD/2PC = PD/2R$ times the area of the sphere.

The distance $DC = R - PD$ is seen to be easier to find. We have that triangle ABC is similar to triangle BDC, hence $DC/BC = BC/AC$ or $DC = R^2/(R+h)$. Hence $PD = R - DC = Rh/(R+h)$ and the visible proportion of the earth is $PD/2R = h/2(R+h)$. Setting this equal to 1/3 gives $h = 2R$, i.e. we must be the earth's diameter away from the earth to see a third of it.

[If we let v be the proportion of the earth visible, then $v = h/2(R+h)$ gives us $h = 2Rv/(1-2v)$. If $v = 1/n$, then $h = 2R/(n-2)$, which gives simple values of h for small n. From the triangle CAB, we find that the latitude b of B is given by $\sin b = R/(R+h) = 1-2v$. This is a simple value only for $v = 1/4$, when $b = 30°$. For our case of $v = 1/3$, we find $b = \arcsin 1/3 = 19.47°$.]

150. HOW TO LOSE A DAY

Once again, our old friend, the International Date Line produces this odd situation. When one crosses it from east to west, the date is advanced. If you just happen to do this at midnight when the date is normally advanced, then your date is advanced two days at once and you completely skip a day.

[By recrossing the Dateline, one can pass through a sequence of dates: 4th, 6th, 5th, 7th in just over 24 hours! At the poles, the idea of date is very unclear which is why it is eliminated from the problem.]

Chapter 11

151. A GRAVE MISUNDERSTANDING?

The explanation lies not in ourselves, but in our stars — or rather in our calendars. From the early middle ages until the adoption of the Gregorian calendar beginning in 1582, the new year began on the 25th of March. Hence dates up to 25 March were considered as part of the previous year. Because of religious differences, England did not adopt the Gregorian calendar until 1751, after the death of young Thomas Lambert.

Because of the difference between English and Continental dates during 1582–1751, one must be cautious with English historical dates that occur on 1 January to 24 March in these years. Indeed one sometimes sees English dates of this period written 1 February $169\frac{1}{2}$, meaning that some people thought it was 1691 while others thought it was 1692. To confuse matters even more, Scotland changed to using 1 January as the beginning of the year in 1600, though it didn't adopt the Gregorian calendar.

I have since seen this problem used in various places, including on a matchbox.

152. WANT A DATE?

Let's see. It's now 2016. One year ago was 2015, 2 years ago was 2014, ... , n years ago was $2016 - n$, so 2,100 years ago was -84, i.e. 84 BC. Hands up, all of you who got this answer! Unfortunately this is wrong — it was 85 BC because there isn't a year 0! When Dionysius Exiguus set out the calendar in the 6th century, zero hadn't become known in Europe and so he omitted to include a year 0. Thus the year before 1 AD is 1 BC.

Incidentally, the European system of numbering floors of a building starts with the ground floor and then the first floor is what Americans call the second floor, etc. That is, the European system has a floor zero at ground level. Floors downward are sometimes numbered −1, −2, etc., so that the distance from the 2nd floor to the −2nd floor is 3 stories in the US but is 4 stories in Europe and the latter is more mathematically sensible.

The lack of a year 0 is well known to historians of astronomy, but not all historians are aware of it — the 2000th birthday of Virgil was celebrated a year early in the earlier part of the 20th century! The lack of a year 0 is also why centuries have to begin at the beginning of the year ab01, not at the beginning of the year ab00.

I wrote this problem when it was still the 20th century and I could ask for the date 2,000 years ago — this older version was numerically a bit simpler.

153. A SHORTER CENTURY

The year 2000 (which is the last year of the 20th century) was a leap year, but the year 2100 is not, so the 20th century is a day longer. [The 21st century will also be lengthened by a few "leapseconds" to compensate for the fact that the earth is running a bit slow, so it will be a bit less than a day shorter.]

154. A CALENDAR ODDITY

First, I trust you have all noticed that February can have five Saturdays if and only if it is a leap year and 1 February is a Saturday. Then the other Saturdays are 8, 15, 22 and 29 February. Since a year is one day longer than 52 weeks, the days of the week move ahead one day every ordinary year and two days every leap year. Hence in the four years to 1996, the days move ahead by five days, which is the same as moving back two days, so 1 February was a Thursday in 1996. Since two is relatively prime to seven, we will get a multiple of seven days only after seven leap years, i.e. 28 years, and so five Saturdays in February will recur in 2020. [If I had asked the question about 1882, there would be the complication that 1900 was not a leap year — but 2000 was a leap year.]

155. UNLUCKY YEARS

This doesn't require much sophisticated mathematics, but the number of unlucky years is enough that one needs to make some observations and to work carefully. First, you probably recall that the sum of the digits of a number is related to 'casting out nines' and some thought about this tells us that two unlucky numbers must differ by a multiple of nine. Now if we examine the unlucky years in the first century of the second millennium, we have: 1039, 1048, 1057, 1066, 1075, 1084, 1093. (Most English people think of 1066 as a lucky year, but of course it was unlucky for Harold!) There is a clear pattern — each unlucky year is just nine more than the previous, because adding nine diminishes the units digit by one and increases the tens digit by one, so the sum of the digits remains unchanged. However, if we add another nine, we get 1102 and the sum of the digits has dropped by nine because of the carry to the hundreds place. Adding more nines leads to 1111, 1120, 1129, and the sum of the digits is back to 13, after 36 years. We can now easily tabulate the behavior through the centuries and total them up for each millennium. The numbers of unlucky years in the 1st, 2nd, ..., 10th millennia are: 75, 73, 69, 63, 55, 45, 36, 28, 21, 15, so the first millennium was the unluckiest and things get better from then on!

The pattern of gaps is a bit trickier. We have seen that the interval between unlucky years tends to be nine, which is the minimal possible interval. We now have a terminology problem in that many people will say that the gap is eight years, i.e. the number of lucky years between two unlucky years. However the spacing between unlucky years turns out to be easier to deal with and I will use 'gap' to mean the spacing. Now at the end of a century, there is a bigger gap. In this millennium, the century gaps start with 1093 to 1129 = 36 years, then continue with 1192 to 1219 = 27 years, then 18, 18, 27, 36, 45, 54, 63 years, the last being from 1840 to 1903. However, we now discover that the gaps at the ends of millennia are much bigger. The gaps at the end of the 1st, 2nd, ... 9th millennia are:

940 to 1039 = 99;	1930 to 2029 = 99;	2920 to 3019 = 99;
3910 to 4009 = 99;	4900 to 5008 = 108;	5800 to 6007 = 207;
6700 to 7006 = 306;	7600 to 8005 = 405;	8500 to 9004 = 504.

The beginning of the first millennium gives some trouble. One might expect the gap would be from −49 to +49, but there are two complications. It's not quite clear what the sum of the digits of a negative number like −49 should be — is it 13 or −13? In order to maintain the connection with casting out nines, it has to be −13. This shows up also in that the difference between −49 and +49 is 98, which is not divisible by 9. The second complication is that there wasn't a year 0, so there are actually only 97 years between −49 and +49. Because of these problems, I think we'd best ignore everything BC and start with year 1, which gives 48 lucky years before the first unlucky one.

The end of the 10th millennium also gives some problems, but not nearly so complicated. We simply have to decide whether we take the gap 9400 to 10039 = 639 years or the gap 9400 to 10000 = 600 years. In either case, it's the biggest gap in the first ten millennia.

156. A LONG MONTH

The first, quibble, answer is SEPTEMBER which requires 9 letters and is the longest of all the names of the months. (The names of the months are now pretty standard, at least in most Western languages. I would be interested to know of languages where September is not the longest month name — I know Polish has different names.)

The second, non-quibble, answer is OCTOBER, which has an extra hour due to the change from Summer (= Daylight Saving) Time to regular time. This depends a bit on where you are — not all countries use Summer Time and not all of them change on the same day (or even month).

157. HOW LONG IS A MONTH?

Would you believe SEVEN? The quibbler's solution is that the names of the months can have 3, 4, 5, 6, 7, 8 or 9 letters. (The average length of a month is then 6 1/3 letters.)

Now would you believe EIGHT? It takes some knowledge to find the real-time solutions. First, there are the obvious lengths of 28, 29, 30 and 31 days. In the UK, Summer Time starts in March and ends in October, so March is one hour shorter than 31 days, while October is one hour longer than 31 days. (In other countries, the change may occur in different months, but this doesn't change the number of possible month lengths, except in countries that do not use Summer Time at all.) Finally, since 1 January

1972, international time-keeping has been based on a system of atomic clocks which are more accurate than the earth's movement. The earth is irregular and a bit slow by this system, by about one second every 400 days and occasional leap-seconds are added, either at the end of December or the end of June, so these months can be one second longer than 31 or 30 days, respectively. This gives us EIGHT possible month lengths. So far, no year has had two leap seconds, and this is not likely to occur in the near future, so at most seven of these lengths can occur in a given year (at least for some time to come).

[My thanks to John Chambers of the Centre for Time and Metrology at the National Physical Laboratory, UK, for some of these details.]

158. TWIN TIMES

One of the twins has traveled around the world, thereby gaining or losing a day compared to the other.

[Surprisingly, the necessity for an International Date Line was recognized in the 14th century! Problems based on it date back to about 1500, before Magellan's men were startled by discovering they had lost a day.]

159. A LONGER MONTH

The solution is to lengthen the 31st of October by traveling westward. If I have an airplane and am not too far from the pole, then I can fly around the world in 24 hours, i.e. I can fly as fast as the sun. If I start at the International Date Line on midnight at the beginning of 31 October and fly west as fast as the sun, I arrive back at the Date Line 24 hours later and it is still midnight at the beginning of 31 October. I then have a 24 hour day at this point and I have made 31 October last for 48 hours, giving an October which has lasted, for me, 32 days and one hour, though stay-at-homes only get 31 days and one hour.

Chapter 12

160. WATCH YOUR CLOCK

Unfortunately, no, they are never at equal angles. If the three hands are at equal angles, the angle between each pair is 120°. First, let us examine the case of the hour and minute hands, and let us measure the angle in minutes of time, so the whole circle is 60 minutes, and a third of a circle, i.e. 120°, is 20 minutes. If the time is M minutes after 12:00, then the hour hand will be at position $M/12$, while the minute hand is at position $M - 60h$, where h is the number of whole hours which have elapsed. Thus we want $M - 60h = M/12 \pm 20$, which gives us $11M/12 = 60h \pm 20$ or $M = 720h/11 \pm 240/11$. [Properly all these last equalities are congruences (mod 60), since the operations may go above or below the range from 0 to 60, but we can allow for this by letting h be the next or previous value.] Now let $S = 60M$ be the time in seconds after 12:00:00, so we see that the hour and minute hands are 20 minutes apart when $S = 43200h/11 \pm 14400/11$.

By the same reasoning, with $M = S/60$ and letting m be the number of whole minutes elapsed, so the second hand is at $S - 60\,m$ and we want it to be at $S/60 \pm 20$, which gives $S = 3600m/59 \pm 1200/59$. Since 11 and 59 have no common factor, our two expressions for S can be equal if and only if each is actually a whole number. Discarding multiples of 11, the first expression for S is a whole number if and only if $3h \pm 1$ is a multiple of 11, and this happens for $h = 4$ or 7. These also make M a whole number, equal to either 240 or 480, i.e. the time is 4:00:00 or 8:00:00. But at both these times, the second hand is alongside the minute hand and not at 120° to it.

[After composing this problem, I found that Pierre Berloquin, in his *The Garden of the Sphinx*, asks when the hands can all be together. In fact this gives further possibilities for my problem, since the hands are then all at 0° from one another. I leave to you to find the solutions in this situation.]

161. THREE CLOCKS

The short answer is 'No'. As stated, the slow and normal clocks will coincide at 12 again after 72 days and all three will hence coincide at that time.

The only way I can understand the solution is to assume the problem has been misprinted as well as mistranslated. If the slow clock loses 12 minutes each day, then the solution makes sense.

162. PANDIGITAL TIMES

Let us give each digit a separate name, so MM:DD:HH:MM:SS = $AB : CD : EF : GH : IJ$. We have $00 \leq AB \leq 12$; $00 \leq CD \leq 31$ (with the upper limit depending on the month); $00 \leq EF \leq 23$; $00 \leq GH \leq 59$; $00 \leq IJ \leq 59$.

A can only be 0 or 1. Suppose $A = 1$. Then B and F can only take on the values 0, 2 and hence must use both values. But then C must be 3 and then there is no possible value remaining for D. Hence, though he didn't say so, Yoshigahara's assumption that $A = 0$ includes all possibilities.

So now assume $A = 0$. Then C must be 1, 2, or 3, while E must be 1 or 2. If $C = 3$, then $D = 1$ and so $E = 2$. But then there are no possible values for F. So C and E must take on the values 1, 2.

We now consider two cases. If $C = 1$, then $E = 2$ and so $F = 3$, giving a date of the form $0B : 1D : 23 : GH : IJ$. Now G and I must take on the values 4, 5 and then B, D, H, J can be any permutation of the values 6, 7, 8, 9. This give us $2 \times 24 = 48$ solutions.

If $C = 2$, then $E = 1$. Now G and I must be two values in $\{3, 4, 5\}$ and B, D, F, H, J can be any permutation of the 5 remaining values. We can choose G, I in 6 ways and permute 5 things in 120 ways, giving 720 solutions, hence 768 in total.

It is now easy to determine that the earliest time is 03:26:17:48:59 and the latest time is 09:28:17:56:43.

163. A VERY ODD CLOCK

The car I had at the time had a digital clock with numbers given by the common seven bar LED display which has three cross bars and four upright bars. One of the light bars had burned out, namely the upper right vertical bar in the second digit. The clock was of the type where 9 uses the bottom bar. [This really did happen!]

164. AN UPSIDE-DOWN TIME

On a seven-bar digital display, the following digits read the same when turned over: 0, 1, 2, 5, 8 and the digits 6 and 9 interchange. However, the first digit can only be 0, 1 or 2, and the third digit can only be 0, 1, 2 or 5. This gives 12 cases, but one is not a proper time: 00:00; 01:10; 02:20; 05:50; 10:01; 11:11; 12:21; 15:51; 20:02; 21:12; 22:22. (The improper time is 25:52.)

165. TIME IN REVERSE

On a seven-bar digital display, the following digits read the same when reversed: 0, 1, 8 and the digits 2 and 5 interchange. However, the first digit can only be 0, 1 or 2, and the third digit can only be 0, 1, 2 or 5. As before, this gives 12 cases, but one is not a proper time. The legal times are: 00:00; 01:10; 02:50; 05:20; 10:01; 11:11; 12:51; 15:21; 20:05; 21:15; 22:55. (The illegal time is 25:25.)

Of these, four times: 00:00; 01:10; 10:01; 11:11 are also the same upside down.

Chapter 13

166. THE SQUASHED FLY

How many people got 50 miles? Hands up! Good, I'm glad to see so many people know this problem. After all, the trains are approaching from 100 miles apart at a total speed of 100 mph, so they'll collide in just one hour, during which time our fly has flown 50 miles (no flies on him!).

Sadly, you're all wrong! Since the first train is going at 60 mph and the poor fly can only do 50 mph, he remains stuck fast on the front of the first locomotive, totally unable to do anything but stare at the oncoming disaster and mutter 'It's a helluva way to run a railroad.'

Moral: don't solve your problem until you've read it.

[Some people claim that the fly would be able to head away from the first locomotive at a total speed of 110 mph, but air resistance would keep him from getting more than a negligible distance.]

167. ROUND AND ROUND AND BACK AGAIN

Let V be the velocity or speed of the cyclist and let v be the velocity of the runner and let $r = V/v$ be their ratio. Suppose the track has length L. Let t be the time when the cyclist overtakes the runner and let T be the time when they both get back to the start. At time t, the cyclist has gone L further than the runner, so $Vt = vt + L$. The cyclist is now at distance vt from the start. So at time T, we have $L = vT$ and $L + 2vt = VT$. Despite having only three equations in five unknowns, we can determine something. Substituting $L = vT$ into the other equations gives $Vt = vt + vT$ and $vT + 2vt = VT$. We rewrite these as: $(V - v)t = vT$ and $2vt = (V - v)T$. Thus $T/t = (V - v)/v = 2v/(V - v)$ and $(V - v)^2 = 2v^2$. Now consider $r = V/v$, the desired ratio of the velocities. Putting this into the last equation and canceling the common factor of v^2 leaves us with $(r-1)^2 = 2$, so $r = 1 + \sqrt{2} = 2.414\ldots.$

Note that we also know $T/t = (V - v)/v = V/v - 1 = r - 1 = \sqrt{2}$. Further, the first meeting point is at vt, which is

$$vt/L = vt/vT = t/T = 1/\sqrt{2} = .707\ldots$$

of the distance round the track. The shape of the track is immaterial to the problem.

168. A KNOTTY NAUGHTY NAUTICAL PROBLEM

I bet you got 2260 hours — but this is not the answer! A knot is a speed of one nautical mile per hour, so a knot per hour is an acceleration. That is, the ship will accelerate from 0 knots to 1 knot in the first hour. It will be going 2 knots at the end of the second hour, 3 knots at the end of the third hour, ... School physics tells us that the distance, s, traveled in time t, is given by $\frac{1}{2}t^2$. So the ship will take $\sqrt{4520} = 67.23$ hours to get to Honolulu. Note that it will be going at 67.23 knots when it gets there, so the surfers had better watch out!

[Inspired by a comment of J. E. Littlewood in *Littlewood's Miscellany*.]

169. LUNATIC GRAVITY

A) A ball thrown straight up from ground level (i.e. $s_0 = 0$) with initial velocity v_0 has velocity $v = -gt + v_0$. At the peak of its travel, $v = 0$, which gives us $t = v_0/g$. Putting this into $s = -\frac{1}{2}gt^2 + v_0t$ gives $s = v_0^2/2g$, so that decreasing g to 1/6 of its value will increase s to 6 times its value and this assertion is true.

B) The ball returns to ground level, i.e. $s = 0$, when $-\frac{1}{2}gt^2 + v_0t = 0$, which gives us $t = 2v_0/g$. Note that this is just twice the time required to get to the top. Again, changing g to 1/6 of its value changes t to 6 times its value, so this assertion is true.

C) If the well has depth d, then the stone hits bottom when $d = \frac{1}{2}gt^2$, so $t = \sqrt{(2d/g)}$ and decreasing g to 1/6 of its value changes t to $\sqrt{6} \approx 2.45$ of its value, so this assertion is false.

D) A horizontally fired bullet travels at constant horizontal speed until it has fallen to the ground. Hence the distance traveled is just the horizontal velocity times the time required to fall from the height of the muzzle to the ground. The previous answer shows that this time is only increased by $\sqrt{6}$ and not by 6, so this assertion is false.

E) However, if the bullet is fired at a 45° angle, then it stays in the air 6 times as long on the moon according to B), hence it does travel 6 times as far.

170. WEIGHT WATCHING

The kilogram of feathers is still heavier than the kilogram of gold in the sense that it will weigh more than the gold if both are weighed in a vacuum. When we weigh things in air, the buoyancy of the air supports the object and causes the scale to register less than the true weight, which would be obtained if we weighed it in a vacuum. Since feathers are much less dense than gold, they are buoyed up more by the air and hence their true weight is greater.

[Properly, a kilogram is a measure of mass, not weight, and the metric system does not even talk about a kilogram weight, but I use it to mean the weight exerted by a kilogram mass weighed in a vacuum, which is about 9.8 newtons at the earth's surface in the metric system. A kilogram mass exerts a weight of one kilogram less the weight of the medium displaced, which is air in our case. Hence a kilogram mass of a less dense material, e.g. feathers, will weigh less, in air, than a kilogram mass of a denser material, e.g. gold. Indeed, since gold is over twice as dense as brass, which is generally used for weights, a kilogram mass of gold will weigh more, in air, than a kilogram weight. So, in air, a kilogram weight of feathers is more than a kilogram of mass, while a kilogram weight of gold is less than a kilogram of mass.]

[Derived from Herbert McKay, *Fun with Mechanics*, 1944.]

An old version of this problem is: Hans Sachs (attrib.); *Useful Table-talk, or Something for all; that is the Happy Thoughts, good and bad, expelling Melancholy and cheering Spirits, of Hilarius Wish-wash, Master-tiler at Kielenhausen.* No publisher, place or cover, 1517, not seen — discussed and quoted in: Sabine Baring-Gould; *Strange Survivals Some Chapters in the History of Man*; (1892), 3rd ed., Methuen, 1905, pp. 220–223. 'In this collection also appears the riddle: "Which is heaviest, a pound of lead or a pound of feathers?" which everyone knows, but with an addition, which is an improvement. After the answer, "Each weighs a pound, and they are equal in weight," the questioner says further: "Not so — try in water. The pound of feathers will float, and the pound of lead will sink."' Of course this is confusing weight and density.]

171. WHEEL TROUBLE

One must look closely at the wheels of a railroad car to see the answer. The wheels are tapered, with the larger part toward the inside of the car, and the rails are slightly rounded. Consequently, when the car goes around a bend, its inertia (generally called centrifugal force) causes it to move a bit outward on the rails. The outer wheel then rides out and up, giving it a larger radius, while the inner wheel moves in and down, giving it a smaller radius.

The diagram shows an exaggerated view of the wheels and axle.

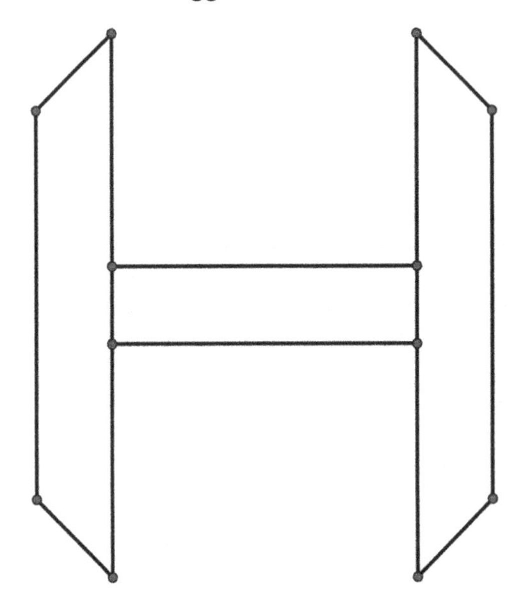

172. SCREWED-UP

The handedness remains the same when it is turned over. You may be able to see this mentally, but I find the following the most convincing argument. Consider putting a nut on a bolt. If the handedness of a helix changed in turning it over, then half the time we tried to put a nut on a bolt, it wouldn't fit and we would have to turn it over. But we know this isn't true — it doesn't matter which end of the nut is up. [As I recall, I heard this from the late Rick Stauduhar when we were students at Berkeley in the 1960s.]

173. RACING ALONG

One can work out the time for each of the flat, uphill and downhill sections separately, but there is a much easier solution. For each mile uphill, the car takes 1/30 hr and for each mile downhill, it takes 1/60 hr. Hence it takes $3/60 = 1/20$ hour to go 2 miles, i.e. its average speed is

$$2/(1/20) = 40 \text{ mph},$$

which is the same as the speed on the level sections. Since the track has as much uphill as downhill, its average speed over the whole course is 40 mph and it takes 1 1/2 hours to complete it. Notice that the arrangement of sections is totally irrelevant!

174. CAUSE FOR REFLECTION

The shortest mirror can be precisely 2 ft 9 in tall, i.e. exactly half her height. It must extend from halfway between her eyes and her toes to halfway between her eyes and the top of her head. Then she can see her whole height from any distance away.

To see this, imagine her mirror image self, who is also 5 ft 6 in high and is standing as far behind the mirror as Jessica is standing in front. The lines of sight from Jessica's eyes to the mirror image's toes and head are bisected by the plane of the mirror, so they are halfway between the eyes and the toes and head at the mirror.

[The width of the mirror is slightly complicated by the fact that we have two eyes, but if we assume one is really dominant, then the width of the mirror can be half the width of the body and it must extend between the points halfway from the dominant eye to the sides.]

175. FURTHER CAUSE FOR REFLECTION

You can see 7 images of yourself. You see 3 ordinary images, one in each single mirror, by looking straight at the mirror. When you look at the 3 edges where two mirrors meet, you see 3 images which are doubly reflected. This causes them to be really reversed. E.g., if you look at the vertical edge in front of you and you move your head to the right, the image moves in the opposite direction — to your left. When you look at the corner where all three mirrors meet, you see an image which has been triply reflected, causing both right and left and up and down to appear reversed. I.e., if you move your head up and right, the image will move down and left (with respect to yourself). The image has your eye at the corner.

176. OVERTAKING VERSUS MEETING

Let the rates be J and S miles per hour. When they run in the same direction, Jessica is ahead by $(J - S)\,t$ at time t. Whenever this is an integer, she is overtaking Stella and she also does this when $t = 1$, so $J - S$ is the total number of overtakings, i.e. $J - S = 3$. When Stella runs in the opposite direction, the distance between them is $(J + S)\,t$ at time t. Whenever this is an integer, they are meeting, which also occurs when $t = 1$, so $J + S$ is the total number of meetings, i.e. $J + S = 11$. Hence $J = 7$ and $S = 4$.

177. WELL TRAINED

Let V be the speed of the passenger train and let v be the speed of the freight train and let D be the distance from London to Newcastle. If they meet at time T hours after starting, then $(V + v)T = D$. We also have that $vT/V = 1$ and $VT/v = 4$. This gives us three equations in four unknowns, so we cannot solve them unless we have some more information. What we want is the ratio VT/D, which is $V/(V + v)$, and we can determine this if we know V/v. But the ratio of the latter two of our equations gives us $V^2/v^2 = 4/1$, so $V/v = 2$ and $VT/D = 2/3$. So the passenger train covers the last 1/3 of its journey in 1 hour, hence it takes 3 hours for the whole trip and the freight train takes 6 hours.

[Note that although we cannot solve the problem completely, we can determine quite a bit.]

178. ANY OLD IRON?

There is the same amount of water before and after and the boat is still floating, so it displaces its weight of water before and after. When the iron is in the boat, it is displacing a volume of water equal to its weight of water, but when it is in the water it displaces only its own volume. Since iron is considerably denser than water, there is more water displaced when the iron is in the boat and the water level in the lock is then higher.

To answer the second question, we need a little notation. Suppose the iron has volume V and density d (relative to the density of water, which can be taken to be 1 by appropriate choice of units). So the iron weighs dV. Then the difference in volume displaced before and after is $dV - V = (d-1)V$. If the lock has area A, then the water level will drop by $(d-1)V/A$. The barge weighs dV less when the iron is dumped over, so it displaces a volume of water less by dV. If the boat has average cross-sectional area a, then it will be dV/a higher in the water. Since $d - 1 < d$ and $A \geq a$, the drop in water level is less than the rise of the boat, so the boat rises with respect to the ground level. [If we replace iron by the same volume of a denser material, then the water level goes down more and the boat is higher in the water and with respect to ground level.]

[A different solution, due to my colleague Laurie Dunn, considers the process as taking two steps. First imagine the iron placed on the ground. Then the boat will rise with respect to the water by an amount equal to the weight divided by the area of the boat, i.e. dV/a, while the water level in the lock drops by dV/A. Since $A \geq a$, the boat rises with respect to the ground. Now put the iron in the water. This raises the water level and the boat goes up with it, so the boat has enjoyed two rises with respect to the ground level.]

179. WATCH THE BOUNCING BALL

In fact Jessica is right — the ball does bounce infinitely many times, but after a few bounces, you can't see it happening. It is sometimes possible to hear quite a number of bounces if you drop a very bouncy ball onto a hard surface. But it doesn't take infinite time, because the time for each bounce gets small sufficiently fast that the total time for all the bounces is finite.

To see this, suppose the ball is dropped from a height H. From school physics, it will reach the ground at time T where $H = \frac{1}{2}gT^2$, g being the acceleration of gravity. Thus $T = \sqrt{(2H/g)}$. School physics also tells us that the time for a ball to bounce up is the same as the time to fall back down. Since the ball goes to half the height, the total time (i.e. up and down) for the second stage is $2\sqrt{(2H/2g)} = 2T/\sqrt{2}$ and at each further stage, the height is divided by 2 resulting in the time being divided by $\sqrt{2}$. Thus the total time will be $T + 2T/\sqrt{2} + 2T/(\sqrt{2})^2 + 2T/(\sqrt{2})^3 + 2T/(\sqrt{2})^4 + \cdots$, which is the same as $T + 2T/\sqrt{2}\ [1 + 1/\sqrt{2} + 1/(\sqrt{2})^2 + 1/(\sqrt{2})^3 + 1/(\sqrt{2})^4 + \cdots]$. The sum in the [ ] is an infinite geometric progression, whose sum is $1/(1 - 1/\sqrt{2}) = \sqrt{2}/(\sqrt{2} - 1) = 2 + \sqrt{2}$. Noting that $2/\sqrt{2} = \sqrt{2}$, we get that the total time for the infinite number of bounces is just $(3 + 2\sqrt{2})T = 5.828\ldots T$, i.e. a bit less than $6T$.

180. JUMPING OVER THE MOON

If we examine the mechanics of the high jump, we find the initial answer is unsatisfactory. I don't know how tall Fosbury is, but let's say $6'$ for a nice round number. When he jumps, he starts vertically and his center of gravity is about $3'$ above the ground. When he crosses over the bar, he is very flattened out horizontally and his center of gravity will be about at the height of the bar. Thus the real work done by Fosbury against gravity consists in raising his center of gravity by $4'4\frac{1}{2}''$, so on the moon he will be able to raise his center of gravity by six times this amount, i.e. $26'3''$. Adding this to the $3'$ height at which his center of gravity starts out, he will be able to jump about $29'\ 3''$.

This question was popular when I was a student in the 1960s. Since then, high-jumping techniques have improved to the point where a high-jumper doesn't actually get his/her center of gravity over the bar. The effect of this is to reduce the value of $29'\ 3''$. If the jumper's center of gravity passes a distance d below the bar, then he/she can clear a bar at height $29'\ 3''$ less $5d$ on the moon. The difference isn't much and is probably less than the error in assuming the center of gravity is at half the person's height.

181. FURTHER REFLECTIONS

When you close one eye looking into a corner mirror, the image shifts so that the line of the corner goes vertically through the open eye. The open eye tends to get completely lost behind the corner line. When you shift to the other eye, the image shifts to having the corner line going through the other eye. Neither single-eye image is complete, but when one uses both eyes, they are overlapped to produce an image which shows your whole self, including both eyes, with a line between the eyes.

182. A CRAFTY WEIGHTLIFT

Basically this is a swiz. The amount the weight lifter can lift is irrelevant. When pulling down on a rope, the most force one can exert is one's weight. The mechanical advantage is correct, so the most our man can lift is a bit less than double his weight!

Chapter 14

183. PATIO PATH PAVING

It is possible to do this by direct trial, but it gets pretty tedious. It is much easier to start by considering shorter paths. Let $P(n)$ be the number of ways to pave a 2 by n path. Clearly $P(1) = 1$ and the problem has already pointed out that $P(2) = 2$.

Consider a 2 by 3 path. I can start out by putting one slab crossways. I am then left with a 2 by 2 path which I can pave in $P(2)$ ways. On the other hand, I can start out by putting a slab longways. Then I have to put another slab longways beside it in order to fill up the width of the path. Thus I now have a 2 by 1 path left over, which I can fill in $P(1)$ ways. So $P(3) = P(2) + P(1) = 2 + 1 = 3$. In the same way, we see that $P(4) = P(3) + P(2) = 3 + 2 = 5$, etc. We can write the general rule as $P(n) = P(n-1) + P(n-2)$, i.e. each term is the sum of the two previous terms. The sequence of terms is thus:

$$1, 2, 3, 5, 8, 13, 21, 34, 55, 89, 144, 233, 377, 610, 987, \ldots .$$

and we want the tenth term in the sequence, namely $P(10) = 89$.

[Our sequence is the famous Fibonacci numbers, introduced by Leonardo of Pisa, known as Fibonacci, in his *Liber Abaci* of 1202, which converted Europe to the Hindu-Arabic numerals. His problem involved the offspring of a pair of rabbits giving rise to the same sequence — $P(n)$ was the total number of rabbits after n months. If you know a bit about Fibonacci numbers, you might like to figure out how many pairs of rabbits there are now, assuming they started at the beginning of 1202.]

184. HALF A CUBE

Let the four cells or cubelets on the bottom level be numbered 1, 2, 3, 4 in cyclic order and let the cells above them be 5, 6, 7, 8, in the same order. We can always turn the cube so that cell 1 is one of the chosen cells, i.e. is red. Let us consider the largest clump (or component) of adjacent chosen or red cells. If there are at least three cells in this clump, then we can bring three of them into positions 1, 2, 3. Then we see that all 5 choices of a fourth cell give different patterns. Now suppose our largest clump has two adjacent cells. We can bring these into positions 1, 2. Now we can't choose 3, 4, 5 or 6 as that gives a larger clump, so the only case here is 1, 2, 7, 8 — two parallel double cubes. Finally, suppose no two cells are adjacent. There is essentially just one such pattern: 1, 3, 6, 8. This gives a total of 7 patterns — much smaller than the 70 claimed!

The patterns 1, 2, 3, 5 and 1, 2, 3, 7 are mirror images, which might be considered the same in some problems, but here we are dealing with a real physical cube so we cannot reflect it.

[Notice that in each pattern, the red cells are congruent to the blue cells. It is also true that every division of the 2×2 square into 2 red and 2 blue cells has the red cells congruent to the blue cells. Does this continue into higher dimensions?]

185. SOLID DOMINOES

No, it cannot be done. View the $3 \times 3 \times 3$ array as a three dimensional chessboard, with the cells alternately colored black and white. Suppose the corners are colored black. Then the layers look like the following.

```
BWB   WBW   BWB
WBW   BWB   WBW
BWB   WBW   BWB
```

There are $5+4+5 = 14$ black cells and $4+5+4 = 13$ white cells. Now when we remove the middle cell, we are removing a white cell and leaving a pattern of 14 black and 12 white cells. Since a domino covers one black and one white cell, no matter where it is placed, no collection of dominoes can cover the board with the middle deleted.

[I discovered this some years ago. Martin Gardner told me it has been discovered previously, but I have never seen it in print.]

186. ALL TIED UP!

There are 13 ways a three-horse race can finish. I list them below, where e.g. A, BC indicates that A came first and then B and C were tied.

ABC;
AB,C; AC,B; BC,A; A,BC; B,AC; C,AB;
A,B,C; A,C,B; B,A,C; B,C,A; C,A,B; C,B,A.

Systematic enumeration gives 75 ways a four-horse race can finish. There is one way with all horses tied, 14 ways with the horses in two groups, 36 ways in three groups (i.e. with a single tie) and 24 ways in four groups (i.e. with no ties).

187. DICING AROUND

There are no less than 16 different standard dice! The six arrangements of spots or pips are as follows.

Of these, the 1, 4 and 5 are the same in all orientations, but the 2, 3 and 6 each have two possible orientations on their face. This gives $2 \times 2 \times 2 = 8$ ways for these three numbers to be oriented. Conveniently, the 2, 3 and 6 faces can all be seen at once since they adjoin one corner of the die.

There are also 2 possible ways to put the numbers on a standard die, as remarked in the problem. We can always turn the die so the 1 is up and the 6 is down. Then we can turn it about the vertical axis so the 2 is front and the 5 is back. But then the 3 and 4 can be either right and left or left and right and these two patterns are mirror images and one cannot be turned into the other by movement in three dimensions. This gives us $8 \times 2 = 16$ dice in all.

Lest anyone think this is a purely hypothetical problem, it only took me about two dozen visits to game, magic and puzzle shops to obtain all 16 of these dice! It is commonly held that a die should have the 1, 2, 3 arranged anticlockwise (or counterclockwise) at a corner, but I have 8 examples of this form and 8 examples with the opposite, mirror image, clockwise arrangement. Most shops have both forms. Ray Bathke, who has the games shop at Camden Lock in London, told me that he had noticed the phenomenon when he supplied dice with games and could not get them all aligned in the same way. He said customers sometimes complained that the dice weren't the same as in previous orders.

In fact, Ray later showed me that there are really 32 dice! Dice from the Far East are often made with the spots of the two running horizontally or vertically on its face rather than diagonally. He already found four different dice with this 'orthogonal' two-pattern, and one of these has the clockwise arrangement. I have since obtained another and I obtained four further types in Beijing in 2004, giving me 9 of the 16 possibilities.

[I recall it was Richard Guy of the University of Calgary who was the first to tell me about the different forms of dice. Ray Bathke has recently shown me oriental dice where the pips are heart-shaped — this vastly increases the number of dice.]

188. COUNTING NUMBERS

With 1, 2, 3, one usually thinks one can form $3 \times 2 \times 1 = 6$ numbers, namely: 123, 132, 213, 231, 312, 321. After thinking a bit, one realizes that Mr. Hammer is also counting 12, 13, 21, 23, 31, 32 and 1, 2, 3. With the 10 digits, one can form 10 one-digit numbers, 10×9 two-digit numbers, $10 \times 9 \times 8$ three-digit numbers, ... , $10 \times 9 \times 8 \times \cdots \times 1$ ten-digit numbers. This gives us $10 + 90 + 720 + 5040 + 30240 + 151200 + 604800 + 1814400 + 3628800 + 3628800 = 9864100$ numbers.

But this includes numbers that start with 0. We can either count those numbers which start with 0, getting $1 + 1 \times 9 + 1 \times 9 \times 8 + \cdots$ or we can count the numbers that don't start with 0, getting $9 + 9 \times 9 + 9 \times 9 \times 8 + \cdots$. More simply, we notice that precisely 1/10 of our numbers start with any given digit, so the numbers that don't start with 0 are 9/10 of our previous total, i.e. 8877690.

189. THE DIE IS CAST

There are 8 corners of a die and the sum of three faces can be the following:

$$1+2+3=6;\quad 1+2+4=7;\quad 1+3+5=9;\quad 1+4+5=10;$$
$$6+2+3=11;\quad 6+2+4=12;\quad 6+3+5=14;\quad 6+4+5=15.$$

When we sum two such values, we get every possible value from 12 through 30. However the probabilities do not behave as simply as for an ordinary table. Here there are $8 \times 8 = 64$ equally likely cases. The number of cases giving a certain value is divided by 64 to give the probability of that value, so we need only look at the number of cases. I list the number of cases that give each value below. Note that symmetry makes V and $42 - V$ have the same number of cases.

12&30	13&29	14&28	15&27	16&26	17&25	18&24	19&23	20&22	21
1	2	1	2	4	4	5	4	5	8

As can be seen, the number of cases does not behave at all simply.
[Below I list the values that occur for each possible number of cases.

Number of cases, N	Values that occur in N cases
1	12, 14, 28, 30
2	13, 15, 27, 29
4	16, 17, 19, 23, 25, 26
5	18, 20, 22, 24
8	21]

[The interested reader may work out what happens on a table that causes edges of the dice to face up.]

190. SEX AND THE HONEYBEE

It is most direct to carefully tabulate the number of ancestors — males $M(i)$, females $F(i)$ and total $T(i)$ — in the i-th generation back. The following relations are seen to hold. $M(i+1) = F(i)$ since only the females have male parents. $F(i+1) = T(i)$, since everyone has a female parent.

	Generation										
	0	1	2	3	4	5	6	7	8	9	10
$M(i)$	0	1	1	2	3	5	8	13	21	34	55
$F(i)$	1	1	2	3	5	8	13	21	34	55	89
$T(i)$	1	2	3	5	8	13	21	34	55	89	144

$T(i) = M(i) + F(i)$, by definition of total. Applying these several times, we see that

$$F(i+2) = T(i+1) = F(i+1) + M(i+1) = F(i+1) + F(i),$$

which is the recurrence for the Fibonacci numbers. Indeed, the sequence $M(i)$ is the standard Fibonacci sequence, while $F(i)$ and $T(i)$ are the same sequence shifted ahead by one and two steps.

191. FERRYING FIVE

Let the travelers be denoted F, D, W, G, C. There are two minimal solutions, with DG or DC crossing first. I will describe the first and leave you to construct the second.

DG cross, D returns. FD cross, F returns. FW cross, D returns. DC cross. This takes 7 crossings, which is the minimum number of crossings even if there are no restrictions on who can stay with whom.

192. ASSORTED VOLUMES

The solution depends on the 'longest increasing subsequence' in the given order. This idea is best explained by an example, say, with 5 books. Let the books be denoted by 1, 2, 3, 4, 5 in order of increasing size and suppose they are in the order 3, 4, 2, 5, 1. A subsequence is any part of these, not necessarily consecutive, considered in the same order, e.g. 3, 4, or 3, 2, 1. Such a subsequence is increasing if the values are in increasing order, e.g. 3, 4 or 3, 4, 5. Consider the longest increasing subsequence of the given order (or one such if there are several). Then take one of the other books out and insert it in its correct place with relation to this subsequence and this increases the length of the subsequence by one. Hence we can put the books into their correct order in a number of steps equal to the number of books less the length of a longest increasing subsequence. Since no move can increase such a subsequence by more than one, there is no way to put the books into order in fewer steps. From this, we see that the worst possible case is when the books are in reversed order and it then takes one step less than the number of books. E.g. 5, 4, 3, 2, 1 requires 4 steps to put into ascending order.

193. DOUBLES MIX-UP

Let the pairs be denoted A, a, B, b, C, c, D, d. Since A and a are never at the same table, we will designate the tables as A and a. On the first night A will partner one of b, c or d, and their opponents will be a man and a woman from the other two pairs. This can occur in $3! = 6$ ways which differ in just the permutation of the names. Consider some one of them, say Ab vs. Cd. Then at the other table, one can have either aB vs. cD or aD vs. Bc which gives us 2 essentially distinct arrangements on the first night. Now we claim that the rest of the arrangement is determined, except that the remaining two nights can be interchanged. On one of these nights, A will play with c, so let us suppose that is the second night. Since A has already played against d, the others at A's table must be Db. The second table has aBCd, but a has already played with B, so this table must have aC vs. Bd. On the third night, there is even less choice and the tables are: Ad vs. Bc and aD vs. bC. The other arrangement on the first night also gives a solution and we summarize them below.

A's table	a's table
Ab vs. Cd	aB vs. cD or aD vs. Bc
Ac vs. bD	aC vs. Bd or aB vs. Cd
Ad vs. Bc	aD vs. bC or aC vs. bD

194. A CHAIN GAME

One's initial reaction is that it will require 8 openings and closings regardless of the lengths. However, if we have a segment of length 1, we can open it and use it to join two other segments. So with 9 segments, if at least four of them have length 1, we can open these four and use them to join the remaining five segments, taking just four times 15 minutes or one hour to do the job. If we have n segments and n is an odd number, then it has the form $2k + 1$ and we can do the job in k openings and closings when there are k segments of length 1. When n is an even number, it has the form $2k$

and we can do the job in k openings and closings when there are $k-1$ segments of length 1.

195. IN A TEARING RAGE

There are three possible answers, depending on how the paper was folded and how it was torn. There might be more if the folds are made diagonally, but this is certainly not a very neat way to fold an ordinary sheet of paper. Let us say the whole sheet has area 1.

Suppose I made the second fold parallel to the first. If I then tear perpendicular to the folds, I get two pieces, each half of the whole sheet. If I tear parallel to the folds, or diagonally, then I get five pieces, two having area 1/8 and three having area 1/4.

It is more normal to make the second fold perpendicular to the first. Then tearing parallel to either fold gives three pieces of areas 1/4, 1/4, 1/2. Tearing diagonally from the original center to the original corners gives four pieces of area 1/4, but tearing along the other diagonal will give five pieces, four of area 1/8 and one of area 1/2.

[Research problem — what if I folded or tore it a different number of times?]

196. NOT SO LIKELY

This is actually a very straightforward exercise in probability. Basically one is taking three cards from the deck, without replacing them. Since we want the probability P of at least one of the events A, 2 or J occurring, it is easier to consider the complementary probability, $1-P$, that none of these events occur. There are 12 desirable cards, hence 40 unwanted cards. The probability, $1-P$, of getting three unwanted cards among three tries, is $40/52 \times 39/51 \times 38/50 = .44706$, so $P = .55294$. This probability doesn't seem very close to 'two out of three', i.e. .66667, to me. At first, I just assumed the author was being a bit careless with his numbers, but while writing this problem up, I wondered what would happen with four cards. Then $1-P = 40/52 \times 39/51 \times 38/50 \times 37/49 = .33758$ and so $P = .66242$, which is more than pretty close to 'two out of three'! So apparently someone forgot how many packs to ask the dupe to cut into! Who's the dupe then?

[In *Knowledge* 1 (3 Feb 1882) 301, there is a second version of the problem, where one is allowed to make three separate cuts, so the probability P of not getting one of the three cards is $(10/13)^3 = .45517$, so $1 - P = .54483$, and this version is a bit fairer than the above version. *Knowledge* 1 (10 Mar 1882) 409 considers the first version of the problem, getting the same result that I have given.]

197. ELECTION SPECIAL

Pretty long! There are 16 characters, since we must include the blank space. If the characters were all different, there would be

$$16! = 16 \times 15 \times \cdots \times 2 \times 1 = 20,922,789,888,000$$

arrangements or permutations of the characters. However one letter is repeated 3 times and three letters are repeated 2 times, so that rearrangements of these letters are not different. This means we have to divide 16! by $3! \times 2! \times 2! \times 2! = 48$, getting $16!/48 = 435,891,456,000$ distinct permutations, hence it will take Jessica and Rachel that many seconds. In a year of 365 1/4 days, there are 31,557,600 seconds, so it will take them 13,812.6 years!

A linguistic purist may point out that an arrangement of letters with the space at the beginning and the same arrangement with the space at the end are really the same. 1/16-th of the above permutations have the space in a given position, in particular 1/16-th have the space at the beginning and 1/16-th have the space at the end. Deleting one of these leaves purists with 15/16 of the above numbers, which purists are welcome to calculate.

198. DOUBLE JUMPING

With 8 counters in a row, there are 16 solutions, but half are the mirror images of the other, which differ only in the order of the last two moves. Move 4 to 7. Move 6 to 2. Trial and error soon shows that these first two moves are unique up to reflection. The situation is now almost symmetric:

1 62 3 - 5 - 47 8 (where 62 indicates we have the two counters 6 and 2 in a pile with 6 on top of 2), and we move (1 to 3 or 3 to 1) and (5 to 8 or 8 to 5), in either order, giving 8 solutions and reflection makes 16.

With 10 counters in a row, begin by moving 4 to 1. This leaves a row of 8 counters (with an unimportant gap) and any of the above solutions can be applied. Likewise for any higher even number, one starts with moving 4 to 1 and using any solution for the previous size. [This result was pointed out to me by Martin Gardner.] There may be other solutions — I haven't looked for them.

With 8 counters in a circle, all first moves are equivalent, so I start by moving 5 to 8. Since we want all the final piles to be on even numbers, only counters in odd locations can move and the only possible move is then moving 1 to 4 and then 7 to 2 and 3 to 6 are forced, but can be in either order.

To get the final piles in consecutive locations, move 5 to 8, 7 to 3, 6 to 1 and 4 to 2.

For 10 counters in a circle, move 7 to 10, 5 to 2, 9 to 4, 3 to 6, 1 to 8. Reversing the last two moves leaves the piles consecutive.

199. AROUND THE ENVELOPE

A path which goes over all the lines once and only once is known as an Eulerian path since Leonhard Euler first studied such paths in 1736. Euler observed that if there is an Eulerian path, then every point or vertex of the diagram has as many lines coming in as going out, so it must have an even number of lines at it, except possibly for the end points. In either of the given diagrams, we see that points 1 and 2 have three lines at them, while the other points have four lines, so points 1 and 2 must be the end points. For convenience, we just consider paths starting at 1 — those starting at 2 will then be just the reversals of those starting at 1. (The diagrams are also symmetric about the vertical midline and one could consider mirror image paths as being the same, which would divide our numbers by 2 since no path is equal to its reflection.)

In either case, the lines D and E are equivalent and interchangeable. So we will only count the paths where line D precedes line E and this again divides our problem by a factor of two.

In the first situation, the pattern is also symmetric by interchanging A with B and F with G and I find the following 11 paths.

ADBCFEG, ADBCGEF, ADEFCBG, ADEFGBC, ADGCBEF, ADGFEBC, AFCBDEG, AFGDEBC; CFABDEG, CFDBAEG, CFDEABG.

Each of these gives 4 distinct paths by applying the two mentioned symmetries, so there are 44 paths from point 1 to point 2 (or 88 paths if we count both directions). I have now checked this with my computer program.

In the second situation, the counting is less easy and my original hand work omitted some cases which my program has identified. I find 22 paths starting with A, 22 starting with B and 16 starting with C, giving 60, each of which gives two distinct paths by interchanging D and E, so there are 120 paths from point 1 to point 2.

200. PLANE COLORING

Consider any three points A, B, C, which form an equilateral triangle of side one inch. All three points must have different colors, so we need at least three colors. Suppose we can do our coloring with just three colors. Now consider the point D such that B, C, D form the other equilateral triangle with side BC.

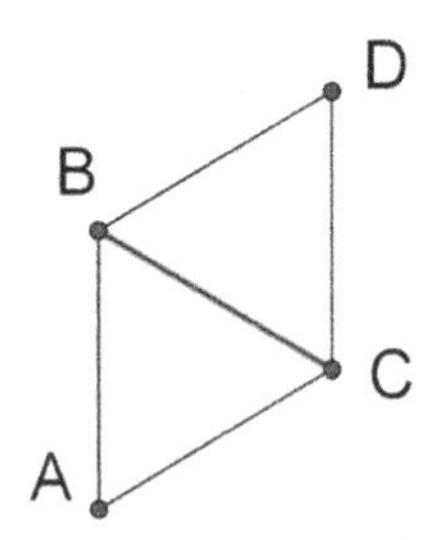

Then D and A must have the same color, namely the color different from B and C. Simple geometry shows that A and D are $\sqrt{3}$ apart. Any two points which are $\sqrt{3}$ apart can be put into this pattern, so any two points which are $\sqrt{3}$ apart must have the same color. Hence the entire circle of points of radius $\sqrt{3}$ about A must have the same color as A. But if D is any point on such a circle, then there is a point one inch away which is also on this circle and we have two points one inch apart with the same color. This contradiction shows that we cannot accomplish our coloring with three colors.

[If one tiles the plane with regular hexagons of diameter a bit less than one inch and colors them with seven colors in the pattern shown, then we have a valid coloring with seven colors.

1 2 3 4 5 6 7 1 2 3 4 5 6 7 1 2 3 4
4 5 6 7 1 2 3 4 5 6 7 1 2 3 4 5 6 7
6 7 1 2 3 4 5 6 7 1 2 3 4 5 6 7 1 2
2 3 4 5 6 7 1 2 3 4 5 6 7 1 2 3 4
4 5 6 7 1 2 3 4 5 6 7 1 2 3 4 5 6

I believe the problem and these two results are due to R. M. Robinson.]

201. A FURTHER CHAIN GAME

By trying a number of cases, one soon notices that if $a_1 \leq n-1$, then one can completely cut the first segment into a_1 open links and use these to join up a_1+1 of the other segments, say the longer ones. This reduces the number of segments by a_1+1, where the one arises because the first segment has been annihilated. We then repeat the argument — if $a_2 \leq n-a_1-2$, we can annihilate the second segment and use it to join up a_2+1 of the other segments. The process continues until the smallest remaining segment has more links than needed to join the remaining segments and then there is no advantage to be gained by opening it all up. The total number of openings and weldings is then $n-1$ less the number of segments which were annihilated. We can phrase this as follows.

Form the sums $a_1+a_2+\cdots+a_j$ until a sum exceeds $n-j-1$. Then the minimal number of openings and weldings is $n-j$.

In the previous Chain Game problem, we got the smallest possible value for $n-j$ by getting the greatest possible value for j, which we did by assuming the first segments had length one, so the above sum had value j.

202. THE BLIND ABBESS AND HER NUNS

By placing 9 in two opposite corners, the Abbess is happy with just 18 people. By placing 9 in each side room, the Abbess is happy with 36 people. Every number from 18 through 36 is possible. To add one person, remove one from a corner and add one to each of the adjacent side rooms.

For any sum S, we can get every number from $2S$ through $4S$.

If all the corner numbers and all the side numbers are equal, then $2A+B=9$, so $A = 0, 1, 2, 3, 4$, and the total number of people is $4A+4B=36-4A=36$, 32, 28, 24, 20. For any sum S, we can get a total of $4S - 4A$, for $A = 0, 1, \ldots, \lfloor S/2 \rfloor$, where $\lfloor x \rfloor$ denotes the greatest integer in x.

To find all the arrangements, I eventually proceeded as follows. Choose A and E such that $0 \leq A \leq S, 0 \leq E \leq S$. Then C and G must both be less than or equal to the smaller of $S - A$ and $S - E$. If we denote this last value by M, there are $(M+1)^2$ ways to choose C and G. Now go through the pairs A, E where $M = S, S - 1, \ldots$. We see there is one case where $M = S$, three cases where $M = S - 1$, five cases where $M = S - 2, \ldots$, so the total number of arrangements is given by: $(S + 1)^2 + 3S^2 + 5(S - 1)^2 + \cdots$. (This can be summed by standard methods to get

$$(S^4 + 6S^3 + 14S^2 + 15S + 6)/6.)$$

For $S = 2$, we get $9 + 3 \times 4 + 5 \times 1 = 26$. For $S = 3$, we get

$$16 + 3 \times 9 + 5 \times 4 + 7 \times 1 = 70.$$

To find the number of arrangements which are not equivalent by symmetries of the square for general S requires more theory than appropriate here. For small S, we can reduce the number of cases by requiring $A \leq E$ and $C \leq G$, but there are still a few cases which have to be examined by hand.

For $S = 2$, there are 9 distinct arrangements whose values of $AECG$ are as follows: 0000, 0001, 0002, 0011, 0012, 0022, 0101, 0111, 1111.

For $S = 3$, there are 19 distinct arrangements, given by $AECG =$ 0000, 0001, 0002, 0003, 0011, 0012, 0013, 0022, 0023, 0033, 0101, 0102, 0111, 0112, 0122, 0211, 1111, 1112, 1122.

203. YARBOROUGH, YAROO!

Consider dealing 13 cards. The probability that the first card is less than a 10 is $32/52 = 8/13$, as Mr. Butler said. But once such a card is received, the probability that the next card is also less than a 10 is diminished to $31/51$, etc. So the probability of a Yarborough is: $32 \times 31 \times \cdots \times 20/52 \times 51 \times \cdots \times 40 =$ $5394/9860459 = .000547$, which is about 1/1828.04, so the correct odds are £1827.04 to £1. Yarborough must have been laughing all the way to his bank!

204. AN UNLIKELY START TO A CHESS GAME

One player can arrange her 8 pawns in $8! = 8 \times 7 \times 6 \times 5 \times 4 \times 3 \times 2 \times 1 = 40320$ ways. But she can arrange her rooks, knights and bishops in two ways each, giving a further factor of 8 for a total of $8 \times 40320 = 322,560$ ways. The second player can also arrange her men in 322,560 ways, so there are $322,560^2 = 104{,}044{,}953{,}600$ ways. But the two players could also turn the board around, making 208,089,907,200 ways. One could conceivably play 10 games a day, so it would take one player 32,256 days $=$ 88.3 years and it would take two players 20,808,990,720 days $= 56,971,911$ years to use up all these ways.

205. WIMBLEDON WORRIES

The possible scores in a set with tie-break are: 6-0, 6-1, 6-2, 6-3, 6-4, 7-5 and 7-6 — just seven possible cases.

For a male loser to win as many games as possible, he must lose three sets by 7-6 and win the other two sets by 0-6, giving him 30 games to the winner's 21. For the women, the scores have to be 7-6 twice and 6-0 once, giving the loser 18 games to the winner's 14.

For the men's game, the winner has to win at least 18 games. There are several ways to have the loser also win 18 games in either four or five sets, e.g. 6-4, 6-4, 0-6, 6-4 and 6-0, 0-6, 0-6, 6-2, 6-4. For the women's game, the winner must win 12 games and the loser can also win 12 games in three sets, e.g. 6-4, 0-6, 6-2. It's not too hard to work out all the possible ways these minima can occur, but it's longer than I can present here.

With tie-breaks, the maximum number of games in a set is 13, so the maximum number of games in a men's match is 65. This is not even, so if both players win the same number of games, the maximum number of games is 64, with 32 for each player. This can be achieved in a number of ways, e.g. 7-6, 6-7, 7-6, 5-7, 7-6, but all ways are just rearrangements of these set scores and there are 12 such rearrangements.

For the women's game, the maximum number of games is 39, so the maximum with both players getting the same is 38 or 19 each, which is easily achieved, e.g. 7-6, 5-7, 7-6, again with all solutions being just rearrangements of these scores and there are just two such rearrangements.

206. DOUBLE JUMPING — 1

For 8 counters, the minimal value is 15 and this is obtained in two ways: 5 to 2, 3 to 7, 1 to 4, 6 to 8 and 5 to 2, 3 to 7, 6 to 8, 1 to 4. These two ways are essentially the same, differing only in the order of the jumps.

For 10 counters, the minimal value is 22, achieved in 4 ways, e.g. by 5 to 2, 7 to 10, 3 to 8, 1 to 4, 6 to 9. The other ways are essentially the same as this solution.

For 12 counters, the minimal value is 31, achieved 6 ways, e.g. by 5 to 2, 9 to 12, 7 to 11, 3 to 8, 1 to 4, 6 to 10. Again the other ways are essentially the same as this solution.

207. DOUBLE JUMPING — 2

It is easy to find a solution for 6 counters: 1 to 4, 2 to 5, 3 to 6. For 8 counters, simply start by moving 4 to 1, which leaves a double pile followed by 6 single counters, so one simply applies the previous solution to the 6 remaining counters. Similarly for any even number $N \geq 8$, we begin by moving 4 to 1 and then apply the solution for $N-2$ counters to the remaining $N-2$ single counters.

For 6 counters, the minimum total of exposed counters is 6 and the unique solution is that given above.

For 8 counters, the minimum value is 11, achieved 3 ways:

1 to 4, 3 to 6, 5 to 8, 2 to 7;

1 to 4, 5 to 8, 2 to 6, 3 to 7;

5 to 8, 1 to 4, 2 to 6, 3 to 7.

The last two solutions are essentially the same, but the first is not so simply related to them.

For 10 counters, the minimum value is 18, achieved 6 ways, e.g. by 1 to 4, 3 to 6, 7 to 10, 5 to 9, 2 to 8. All the other solutions are related to this one.

208. SINGLE JUMPING

A solution for 4 counters is: 1 to 3, 2 to 4. For any even number $N \geq 6$ of counters, begin by moving 3 to 1 and then apply the solution for $N-2$ counters to the $N-2$ single piles remaining.

For 4 counters, the minimum total of exposed counters is 3 and the unique solution is that given above.

For 6 counters, the minimum value is 7, which occurs in two essentially equal ways: 1 to 3, 4 to 6, 2 to 5 and 4 to 6, 1 to 3, 2 to 5.

For 8 counters, the minimum value is 13, which occurs in three essentially equivalent ways, one of which is: 1 to 3, 6 to 8, 4 to 7, 2 to 5.

For 10 counters, the minimum value is 21, which occurs in four essentially equivalent ways, one of which is: 1 to 3, 8 to 10, 6 to 9, 4 to 7, 2 to 5.

209. STAR BRIGHT

I get 58 triangles. Let the side of the little triangles be one. Then there are 24 triangles of side 1, 18 of side 2, 8 of side 3, 6 of side 4 and 2 of side 6.

210. ANOTHER TRIANGLE NUMBER

I find 27 triangles. Let the intersection of Aa and bc be e, of Aa and bd be f and of bd and ac be g. Then I find the triangles: ABC, ABa, ACa, Aab, Aac, Abc, Abe, Abf, Ace, Bac, Bcd, Cab, Cbd, abc, abd, abe, abf, abg, acd, ace, adf, adg, afg, bcd, bcg, bef, cdg. If you find more, please let me know.

211. PICTURE CARD PROBABILITY

When you cut a deck of cards in the ordinary way, you are moving part of the deck from the top to the bottom. However, if you view the deck cyclically, the cycle of cards does not change, only the location in the cycle of the top card changes. So the twelve picture cards always remain in sequence, and they will no longer be together if and only if the top card is one of the last eleven of the twelve. Since all 52 cards are equally likely places for the cut at each stage, the probability of the top card being one of the last eleven pictures is constant, being 11/52. Once the first cut is made, the location of the top card is random and all succeeding cuts have the probability of $11/52 = 0.212$. This is considerably larger than $1/500 = 0.002$. Perhaps our author saw a value of 0.2 and thought it was 0.2%.

Remark: In practice, magicians know that most people cut the deck near the middle and this can be used to make the trick work if the magician places the twelve cards at one end of the deck and just one cut is made. However, after two or three cuts, the location of the top card becomes pretty random.

212. PAINTED CUBES

The unpainted cubes form an inner block of dimensions $A-2$ by $B-2$ by $C-2$, so the question is asking for solutions of

(*) $(A-2)(B-2)(C-2) = ABC/2$.

Before looking for solutions, observe that this is the same as

(**) $(1-2/A)(1-2/B)(1-2/C) = 1/2$. As A, B, C increase, the factors increase. So if $A \leq B \leq C$, then the product of the factors is $\geq (1-2/A)^3$ and this exceeds 1/2 for $A \geq 10$. Hence any solution with $A \leq B \leq C$ has $A \leq 9$. But $A = 3$ or $A = 4$ already makes the factor $1-2/A$ less than or equal to 1/2 and multiplying this by further factors less than one cannot give a solution. Hence we can only have $A = 5, 6, 7, 8, 9$.

I don't see any approach simpler than trying each value of A. I illustrate with $A = 5$. Then (*) becomes $3(B-2)(C-2) = 5BC/2$ or

$$6(B-2)(C-2) = 5BC$$

or

$$6BC - 12B - 12C + 24 = 5BC$$

or

$$BC - 12B - 12C + 24 = 0.$$

Using a generalized form of completing the square gives us

$$BC - 12B - 12C + 144 = 120$$

or $(B-12)(C-12) = 120$. The solutions are then determined from the factorizations of 120 and we find 8 solutions: $A, B, C =$ $5, 13, 132$; $5, 14, 72$; $5, 15, 52$; $5, 16, 42$; $5, 17, 36$; $5, 18, 32$; $5, 20, 27$; $5, 22, 24$.

The case $A = 6$ leads to $(B-8)(C-8) = 48$ and 5 solutions:

$$6, 9, 56;\quad 6, 10, 32;\quad 6, 11, 24;\quad 6, 12, 20;\quad 6, 14, 16.$$

The case $A = 7$ leads to $3BC - 20B - 20C + 40 = 0$ and to factorize this we first have to multiply by 3 to get $9BC - 60B - 60C + 120 = 0$

which gives us $(3B - 20)(3C - 20) = 280$. Now not every factorization of 280 gives integral values for B and C, but we get 4 solutions:

$$7, 7, 100; \quad 7, 8, 30; \quad 7, 9, 20; \quad 7, 10, 16.$$

The case $A = 8$ leads to $(B-6)(C-6) = 24$ and the following 3 new solutions: 8, 8, 18; 8, 9, 14; 8, 10, 12.

The case $A = 9$ leads to $(5B - 28)(5C - 28) = 504$ which gives no new solutions.

This gives 20 solutions. The first one listed has the largest volume and the last has the smallest volume.

213. A PLACING PROBLEM

Let the board be labeled as at the right and let the colors be denoted a, b, c. Looking at a corner, such as A, we see that its neighbors, such as B, D must have the other two colors. The choice of colors is rather arbitrary, so let us fix the colors at the first corner now and we will look at permuting the colors later. So suppose $A = a$, $B = b$, $D = c$, where I am using $A = a$ to mean that cell A has color a, etc.

$$\begin{matrix} A & B & C \\ D & E & F \\ G & H & I \end{matrix}$$

Now an adjacent corner triple, like B, C, F must also have all three colors, and since $B = b$ is already fixed, we must have $C, F = a, c$ or c, a.

In the first case, looking at the lower corner triples, we see that G, H and H, I must both contain both colors a and b. Since we have one a and two bs left, this can only occur if $G, H, I = b, a, b$ and $E = c$.

In the second case, we see that G, H must be colored with a and b, while H, I must be colored with b and c. This can only happen when $G, H, I = a, b, c$ and $E = b$.

So we see there are two types of solution. One has a solidly colored horizontal midline while the other has a solidly colored vertical midline. In each case, the colors can be permuted in $3 \times 2 \times 1 = 6$ ways, so there are 12 solutions. If we ignore the color changes, we have just our two types, but one type is obtained from the other by rotation and color change and so we could say that all the solutions are easily derived from just one basic solution.

[The question of how many distinct solutions there are in such problems is complicated. We have two kinds of symmetry acting: the rotations and reflections of the board and the permutations of the colors. Dealing with one or the other of these is easy, but they interact in a complex manner which makes the counting difficult. In our problem, if we only consider symmetries of the board, there are three types of solution, determined by the color of the solid midline. So there are four different numbers of solutions: 12, 2, 3, 1, depending on who's counting.]

214. HOW MANY CHESS PLAYERS?

The key to this problem is to realize that in a class of a players, there will be $a(a-1)/2$ games played. These are the well-known 'triangle numbers': 1, 3, 6, 10, 15, 21, 28, 36, 45, 55, 66, 78, 91, 105, ... The problem then requires two of these which add up to 100 and there is only one way this occurs: $45 + 55$; which occurs for classes of 10 and 11 players, so there must have been 21 players in total.

[Normally, we let $T(k) = k(k+1)/2$ be the k-th triangle number as it is the number of points in a triangle with k points along each edge. The problem leads to interesting questions about when an integer is a sum of triangle numbers. Fermat asserted and Gauss proved that every number is a sum of three triangle numbers, but this requires including zero as a triangle number. What is the smallest number that is a sum of two positive triangle numbers in two different ways? In three different ways?]

Chapter 15

215. MATCHSTICK WORDS

Here are four seven-letter words: EMANATE, FIFTEEN, VILLAIN, ITALIAN. There is a ten-letter word: INFINITELY.

This was based on a problem I saw in a book somewhere — I recall it was 1930s or 1940s, but I couldn't recall where I saw it. I'm afraid I didn't try to find longer examples — or if I did so, I did it most INATTENTIVELY. This produced more correspondence than any other Brain Twister that I set — perhaps it's a good idea to make a mistake sometimes. 31 correspondents sent lists ranging from 1 to 103 examples and I did a dictionary search to turn up many more. We found four 14 letter words: Alimentatively, Antifemininity, Antifemininely, Inflammatively; and an 18-letter word: Antialimentatively.

In 2005, I found the problem in "Zodiastar" [pseud. of ??]; *Fun with Matches and Matchboxes Puzzles, Games, Tricks, Stunts, Etc.*; Universal, no date [1941 — according to British Library Catalogue], pp. 42–43: A Spelling Bee with Matches. I have long owned two other versions of this book, with variant titles but identical content. He gives the 7-letter examples and the 10-letter example which I gave in the problem, so it seems clear that one of these books was the then forgotten source of the above problem. I am unable to determine who "Zodiastar" was, and it might be a generic name for Universal's anonymous books.

216. A JOLLY GATHERING

UNDERGROUND, which is where troglodytes live. SUPERGROUPS, which is a mathematical term, but also describes the jolly meetings of the various Much Puzzling societies!

217. AN INTERMEDIATE PROBLEM

'And' is between up and down.

Four and five are nine.

218. CALCULATED WORDS

The following are the best that I have found so far. Recall the permissible letters are: B, E, G, H, I, L, O, S, Z.

7 letters: Besiege. Bezzles. Bobbies. Bobbles. Boggles. Bolshie. Boobies. Eggless. Ego-less. Elegies, Elegise. Elogise. Giggles. Globose. Glosses. Gobbles. Goggles. Googles. Heigh-ho. Helixes. Hellish. Higgles. Highish. Hobbies. Hobbish. Hobbles. Hoggish. Hooshes. Iglooes. Legible. Legless. Lessees. Obelise. Obligee. Obliges. Seghols. Sizzles. Seizers. Sessile. Sleighs. Sloshes. Soboles. Zoozoo.

8 letters: Besieges. Blessees (those who are blessed?). Blissless. Eggshell. Elegises. Elogises. Eligible. Ghillies. Gigoloes. Goloshes. Goose-egg. Heelless. Heliosis. Hell-hole. Highheel. Hill-less. Hissless. Hole-less. Isohelic. Obligees. Obsesses. Oologise. Shell-egg. Shoebill. Shoeless. Soilless.

9 letters: Bob-sleigh. Eggshells. Eligibles. Geologise. Globeless. Goose-eggs. Heigh-hoes. Highheels. Illegible. Liegeless. Oologises. Shellhole. Shell-less.

10 letters: Bob-sleighs. Geologises. Goloe-shoes. Shellholes. Sleighbell.

11 letters: Sleighbells. [C. R. Charlton sent in Hillbillies.]

12 letters: Eggshell-less. Glossologise. Highheel-less.

13 letters: Glossologises.

If one also permits the letter X, one doesn't gain much. I only found the following.

7 letters: Hell-box. Sexless.

8 letters: Exegesis. Exigible. Loose-box.

David Vincent sent Googolgoogolgoogol.....

219. UNNUMBERED LETTERS

The dictionary I consulted gives number names up to vigintillion, which is 10^{63} in the American system and 10^{120} in the older British system.

A first occurs in 100, if you say 'a hundred'; or in 101, if you say 'one hundred and one'; or in 1000, if you don't indulge in either of the previous locutions. Other letters are more straightforward.

B — billion ($10^9/10^{12}$);
C — octillion ($10^{27}/10^{48}$);
D — hundred;
E — one;
F — four;
G — eight;
H — three;
I — five;
L — eleven;
M — million ($10^6/10^6$);
N — one;
O — one;
P — septillion ($10^{24}/10^{42}$);
Q — quadrillion ($10^{15}/10^{24}$);
R — three;
S — six;
T — two;
U — four;
V — five;
W — two;
X — six;
Y — twenty.

So the last letter to occur is C.

The letters J, K, Z never occur in the names in my dictionary and since the names of the large numbers are based on Latin roots and J, K, Z are not in the Latin alphabet, it is unlikely they will ever be used, except in colloquial phrases like 'a zillion'. Note that I specified positive whole numbers — zero contains Z and j is often used instead of i as the imaginary unit.

220. ALOHA

A bit of random thinking and some browsing through a dictionary shows there are a number of words of ten or more letters.

10 LETTERS: Anopheline, Apollonian, Panamanian, phenomenal, phenomenon, Philippino, pneumonial,

11 LETTERS: philhellene,

12 LETTERS: nonwholemeal,

13 LETTERS: epiphenomenal,

15 LETTERS: unepiphenomenal,

221. A VERY SNEAKY SEQUENCE

Use $1 = A, 2 = B, \ldots$ and the numbers translate to ONE, TWO, THREE, FOUR, FIVE, ..., so the next word is SIX whose number is 19924.

[R. Kerry & J. Rickard. Problems Drive 1983. *Eureka* 43 (Easter 1983) 11–13 & 62–63, Prob. 3 (i).]